The Doubly Green Revolution

Food for All in the Twenty-First Century

GORDON CONWAY

With Forewords by Vernon Ruttan and
Ismail Serageldin

D0965368

Comstock Publishing Associates a division of
Cornell University Press | Ithaca, New York

To Susan

Copyright © 1997 by Gordon Conway
Vernon Ruttan foreword copyright © 1998 by Cornell University

All rights reserved. Except for brief quotations in a review,
this book, or parts thereof, must not be reproduced in any
form without permission in writing from the publisher.
For information, address Cornell University Press,
Sage House, 512 East State Street, Ithaca, New York 14850.

First published by Penguin Books, 1997
Cornell Paperbacks edition with foreword by Vernon Ruttan, 1998.

Librarians: A CIP catalog for this book is available
from the Library of Congress.

Printed in the United States of America

ISBN-13: 978-0-8014-8610-4 (pbk. : alk. paper)
ISBN-10: 0-8014-8610-6 (pbk. : alk. paper)

*Cornell University Press strives to use environmentally
responsible suppliers and materials to the fullest extent possible in
the publishing of its books. Such materials include vegetable-based,
low-VOC inks and acid-free papers that are recycled,
totally chlorine-free, or partly composed of nonwood fibers. For
further information, visit our website at www.cornellpress.cornell.edu.*

Paperback printing   10 9 8 7 6 5 4 3

Contents

Foreword to Cornell Paperbacks Edition

Gordon Conway poses an exceedingly difficult challenge. The "doubly green" revolution needs to provide the world's farmers with the means to meet the increasing demands of society, while also addressing the economic concerns of the producers and the environmental impact of the technologies. Although societal needs continue to be as great as they were in the past, the sources for potential growth and improvements in agricultural production in the future are not as apparent as they were in the early 1960s as the "green revolution" evolved.

The achievement of the first green revolution was increased food production, which has more than kept pace with the growing demands associated with population and income growth in most areas of the developing world over the last two generations. This improvement in agricultural production can be linked to the proliferation of irrigation, the more intensive use of fertilizer and plant protection chemicals, along with the development of new crop varieties. These simultaneous developments resulted in cropping systems that are capable of responding to higher levels of inputs and management.

The higher crop yields attained during the first green revolution were most readily achieved in robust soil areas. In those areas where agricultural production expanded most rapidly, the rural poor benefited; but, overall, it was the urban poor who were clearly the beneficiaries of the green revolution. Conway now calls for technologies that will enhance the productivity and the incomes of farmers living in poorer soil and resource locations. Areas bypassed by the green revolution have not only failed to realize gains from higher yields, they have also experienced price declines for their commodities on the local and world market. This doubly green revolution must provide greater benefits to the world's poor.

Conway also proposes the doubly green revolution will be more "green." The intensive agricultural production associated with the green revolution has left a negative impact on human health and the environment. For example, pesticide poisoning and nitrates in the drinking water have proved to be detrimental to human and animal health. Damaging environmental effects include increased salinization of irrigated land, nitrous oxides and methane contributing to global climate change, and the resurgence of pests and pathogens resistant to available control methods. Some of these negative spillovers eventually have a depressing effect on agricultural production, the very thing they were intended to improve.

The Doubly Green Revolution is a remarkably useful book in which Conway explores why, and how, this new revolution must go beyond simply responding to aggregate food demand. This book is the place to start for those seeking to understand agricultural development in poor countries during the era of the first green revolution and the challenges to sustainable development over the next several decades.

Vernon W. Ruttan
Regents Professor
Department of Applied Economics
University of Minnesota

Foreword to UK Edition

I am delighted that Gordon Conway has decided to present to a wide audience the case for a new and 'doubly green' revolution which he initially presented to the Consultative Group on International Agricultural Research (CGIAR).

The original 'green revolution' was a splendid achievement. The use by farmers of new, research-based technologies transformed agriculture and created food abundance, primarily in Asia and Latin America. Millions were fed and the very real threat of famine was thwarted. Despite these achievements, the food-security agenda has not yet been completed.

A second and more widespread transformation of agriculture is required to combat the nexus of problems associated with poverty, hunger and environmental degradation. To cope with this reality, the CGIAR launched an eighteen-month renewal programme (May 1994/October 1995) to clarify its visions, refocus its research agenda, improve its governance and operations, and secure stable financial support for its mission. As part of this effort, the CGIAR invited a small international team led by Gordon Conway to propose a new vision for the CGIAR. Their report, entitled *Sustainable Agriculture for a Food Secure World*, emphasized the need for agricultural transformation to be 'doubly green' – with equal importance for productivity and natural resource management – and for the knowledge of farming communities to be respected when research agendas are defined and implemented. Gordon Conway was among the first to argue that the new goals could not be met without a genuine partnership among all components of the global agricultural research system. A Global Forum held under CGIAR auspices last year was a catalyst for this new partnership mode, and a step towards enhanced cooperation among all sections of the agricultural community.

The Conway Report, as it is widely known, was endorsed at a CGIAR ministerial-level meeting held in Lucerne in 1995, and continues to inspire the CGIAR. This book, based on the report, builds on the initial arguments and takes the issues forward to a wider community of readers. Access to food is a human right, and food security is the key to a prosperous and stable world. Equally, protecting the natural resources base on which food productivity depends is an obligation of all of us on Planet Earth. I strongly commend this book to all concerned with ensuring that the human family achieves these goals.

Ismail Serageldin
Chair, Consultative Group on International
Agricultural Research (CGIAR)
and Vice-President, Environmentally Sustainable
Development, World Bank

Acknowledgements

This book draws on a wide variety of sources, but I have given prominence to the work of those with whom I have been associated over the past thirty-five years. Many are named in the text. They include: postgraduate students and faculty at the Imperial College of Science and Technology, the University of Sussex, the Universities of Chiang Mai and Khon Kaen in Thailand, Padjajaran University, Indonesia and the University of the Philippines, Los Banos, Philippines; colleagues at institutes where I have worked or been a director – the International Institute for Environment and Development (IIED), the Institute for Development Studies (IDS), the International Food Policy Research Institute (IFPRI) and the Ford Foundation – and workers in numerous non-governmental organizations, including the Aga Khan Rural Support Programme (AKRSP) in Pakistan and India, Action Aid and MYRADA in India, Winrock International in Nepal, and the Ethiopian Red Cross; and the many scientists at the International Agricultural Research Centres.

I have drawn freely on their writings and am grateful for their advice, experience, and friendship. I am particularly indebted to Robert Chambers, Peter Hazell, Michael Lipton, Simon Maxwell and Ian Scoones, who have read and commented on parts of the book.

The conception of a Doubly Green Revolution was an outcome of the deliberations of a small panel, which I chaired, commissioned to develop a vision statement for the Consultative Group on International Agricultural Research (CGIAR) – the organization that supports the international research centres which, over the past thirty years, have spearheaded the Green Revolution.★ It was presented and adopted at a meeting of ministers of overseas development from the developed countries and of agriculture and natural resources from the developing countries, held in Lucerne in

★ G. R. Conway, U. Lele, J. Peacock and M. Piñeiro, *Sustainable Agriculture for a Food Secure World*, Washington DC, Consultative Group on International Agricultural Research/ Stockholm, Swedish Agency for Research Cooperation with Developing Countries, 1994.

February of 1995. My colleagues on the panel were Uma Lele, of the University of Florida (now at the World Bank), Martin Piñeiro (formerly director of the Inter-American Institute for Cooperation on Agriculture (IICA)), Jim Peacock, Chief of the Division of Plant Industry of Australia's Commonwealth Scientific and Industrial Research Organization (CSIRO), Selçuk Özgediz of the World Bank, Johan Holmberg of the Swedish Agency for Research Cooperation with Developing Countries (SAREC) (now Director, Natural Resources and Environment, Swedish International Development Cooperation Agency (Sida)), Henri Carsalade of the French Centre for International Cooperation in Development-Orientated Agricultural Research (CIRAD) (now Assistant Director-General of the Department for Sustainable Development at FAO), Michel Griffon of CIRAD and Peter Hazell of IFPRI. I am grateful to all of them, to Paul Egger of the Swiss Development Corporation and Robert Herdt of the Rockefeller Foundation, both of the CGIAR's Oversight Committee, and to Ismail Serageldin, chair of CGIAR, for their collegial support and encouragement. Many of the ideas in this book are theirs.

Finally, I would like to thank Khun Akhorn and Khun Chompunute Hoontrakul for providing an idyllic writing environment at Tongsai Bay, Ko Samui, and Mrs Elizabeth Ford, my personal assistant, for her invaluable assistance.

For permission to reproduce copyright material in this book the author and publishers gratefully acknowledge the following:

Figure 1.4: from *World Development Report* 1993 by The World Bank, © 1993 by The World Bank used by permission of Oxford University Press, Inc; Figure 1.8: from *Investing in Food Security*. Briefing paper for the World Food Summit (FAO, Rome, 1986); Figure 2.1: in Bongaarts, J. 1995, *Global and Regional Population Projections to 2025*. In Islam, N. (ed.), Population and Food in the Early 21st Century: Meeting Future Food Needs of an Increasing Population (IFPRI, Washington D.C.), by permission of IFPRI; Figure 2.2: from Conway, G.R. and Barbier, E.R. 1990, After the Green Revolution: Sustainable Agriculture for Development (Kogan Page Ltd/Earthscan Publications); Figure 3.1: *Growth Rates of Per Capita Agricultural and Nonagricultural GDP, Various Asian Countries, 1960–1986*, from World Development Report 1990 by World Bank, © 1990 by the

World Bank, used by permission of Oxford University Press, Inc; Figure 5.1: from *World Development Report*, 1990 by World Bank. © 1990 by The World Bank, used by permission of Oxford University Press, Inc.; Figure 6.2: from Loevinsohn, M.E. 1987, Insecticide use and increased mobility in rural central Luzon, Philippines (*The Lancet* 13 June 1987). © The Lancet Ltd, by permission of The Lancet Ltd; Figure 6.6: from Conway, G.R. and Pretty, J.N. 1991, Unwelcome Harvest: Agriculture and Pollution (Kogan Page Ltd/Earthscan Publications); Figure 6.7 and Figure 6.8: from Houghton et al. 1996, Climate Change 1995: The Science of Climate Change (Cambridge University Press), by permission of Dr N. Sundararaman; Figure 7.5: from Bumb, B.L. and Basnante, C.A. 1996, *The Role of Nitrogen Fertilizer in Sustaining Food Security and Protecting the Environment*. Food, Agriculture and Environment Discussion Paper 17 (IFPRI, Washington D.C.), by permission of IFPRI; Figure 7.14: reprinted from Field Crops Research 5 (1980), pp 201–216, with kind permission from Elsevier Science – NL, Sara Burgerhartstraat 25, 1055 KV Amsterdam, The Netherlands; Figure 8.1 from IRRI 1996 Bt Rice: Research and Policy Issues. IRRI Information Series No. 5 (IRRI, Los Banos); Figure 8.2: from Larkin, P. (ed.) 1994, Genes at Work: Biotechnology (CSIRO), by permission of CSIRO; Figure 8.3: from IRRI 1996 Bt Rice: Research and Policy Issues. IRRI Information Series No. 5 (IRRI, Los Banos): Figure 9.1 and Figure 9.2: from Conway, G.R. 1987, The Properties of Agroecosystems (*Agricultural Systems, 24 pp 95–117*), by kind permission of Elsevier Science Ltd, The Boulevard, Langford Lane, Kidlington OX5 1GB, UK; Figure 11.1: from Conway, G.R. 1972, *Ecological aspects of pest control in Malaysia*. In Farvar M.T. and Milton J.R. (eds.) Careless Technology. © 1972 by the Conservation Foundation and The Center for the Biology of Natural Systems (Washington University), used by permission of Doubleday, a division of Bantam Doubleday Dell Publishing Group Inc., and Tom Stacey Limited, London; Figure 11.2: from Loevinsohn, M.E. Litsinger, J.A. and Heinrichs, E.A. 1988, *Rice insect pests and agricultural change*. In Harris, M.K. and Rogers, C.A. (eds.) The Entomology of Indigenous and Naturalized Systems in Agriculture (Westview Press); Figure 11.3: from Georghiou, G.P. 1986, *The Magnitude of the problem*. In Pesticide Resistance: Strategies and tactics for management. © National Academy of Sciences, reprinted courtesy of the National Academy Press, Washington, D.C.; Figure 12.1: from Conway, G.R. and Pretty, J.N. 1991, Unwelcome Harvest: Agriculture and Pollution (Kogan

ACKNOWLEDGEMENTS

Page Ltd/Earthscan Publications), modified from Figure 13.2 from Brady/Weill, Nature and Properties of Soil (11th edition). © 1974, adapted by permission of Prentice-Hall, Inc., Upper Saddle River, NJ; Figure 12.4: from Preston, T.R. and Leng, R.A. 1994, *Agricultural technology transfer: Perspectives and case studies involving livestock*. In Anderson, J.A. (ed.) Agricultural Technology: Policy Issues for the international community (CAB International, Wallingford), by permission of CAB International; Figure 12.5: from Sanchez, P. 1994, *Alternatives to slash and burn: a pragmatic approach for mitigating tropical deforestation*. In Anderson, op. cit, by permission of CAB International; Figure 14.2, Figure 14.4 and Figure 14.5: from Behnke, R.H., Scoones, I. and Kerven, C. (eds.) 1993, Range Ecology and Disequilibrium (Overseas Development Institute, London), by permission of the publisher; Figure 14.9: from Maddox, B. 1994, Fleets fight in over-fished waters (*Financial Times* 30 August 1994), by permission of Financial Times Graphics; Figure 15.7: from Davies, Adaptable Livelihoods (Macmillan), by permission of the publishers.

Every effort has been made to obtain permission from all copyright holders whose material is included in the book, but in some cases this has not proved possible. The publishers therefore wish to thank all copyright holders who are included without acknowledgement. Penguin UK apologises for any errors or omissions in the above list and would be grateful to be notified of any corrections that should be incorporated in the next edition of this volume.

Editorial Note

Regions. In this book the regional groupings follow those used by the Food and Agriculture Organization, although where data are acquired from other sources there may be some variations, as indicated.

The countries of the *less developed* world ('*the Developing Countries*') are grouped as follows:

East Asia: Brunei, Cambodia, China, Hong Kong, Indonesia, Korea (Democratic People's Republic of), Korea (Republic of), Laos, Macau, Malaysia, Mongolia, Myanmar, Philippines, Singapore, Thailand, Vietnam (sometimes Taiwan). Often included are Fiji, Papua New Guinea, Solomon Islands, Tuvalu, Vanuatu and other developing island countries located in the Pacific Ocean (Oceania less Australia and New Zealand).

South Asia: Afghanistan, Bangladesh, Bhutan, India, Maldives, Nepal, Pakistan, Sri Lanka (sometimes Afghanistan is included under West Asia/North Africa).

West Asia/North Africa: Algeria, Bahrain, Cyprus, Egypt, Gaza Strip, Iran, Iraq, Jordan, Kuwait, Lebanon, Libya, Morocco, Oman, Qatar, Saudi Arabia, Syria, Tunisia, Turkey (sometimes), United Arab Emirates, Yemen.

Sub-Saharan Africa: Angola, Benin, Botswana, Burkina Faso, Burundi, Cameroon, Cape Verde, Central African Republic, Chad, Comoros, Congo, Côte d'Ivoire, Djibouti, Equatorial Guinea, Eritrea, Ethiopia, Gabon, Gambia, Ghana, Guinea, Guinea-Bissau, Kenya, Lesotho, Liberia, Madagascar, Malawi, Mali, Mauritania, Mauritius, Mozambique, Namibia, Niger, Réunion, Rwanda, St Helena, São Tomé e Príncipe, Senegal, Seychelles, Sierra Leone, Somalia, South Africa, Sudan, Swaziland, Tanzania, Togo, Uganda, Western Sahara, Zaire (Democratic Republic of Congo), Zambia, Zimbabwe (sometimes South Africa is not included).

Latin America and Caribbean: Argentina, Bolivia, Brazil, Chile, Colombia, Costa Rica, Cuba, Dominican Republic, Ecuador, El Salvador, French Guiana, Guatemala, Guyana, Haiti, Honduras, Jamaica, Mexico, Nicaragua, Panama, Paraguay, Peru, Suriname, Trinidad and Tobago, Uruguay, Venezuela and the smaller islands of the Caribbean.

The countries of the *more developed* world ('*the Developed Countries*') are grouped as follows:

Eastern Europe/Former Soviet Union (FSU): Albania, Belarus, Bosnia and Herzegovina, Bulgaria, Croatia, Czech Republic, Estonia, Hungary, Latvia, Lithuania, Macedonia, Moldova, Poland, Romania, Russia, Slovakia, Ukraine, Yugoslavia, and the Central Asian Republics (Armenia, Azerbaijan, Georgia, Kazakhstan, Kyrgyzstan, Tajikistan, Turkmenistan, Uzbekistan). In future, the Central Asian Republics will be included in official statistics as part of the developing countries, either as a new separate region of Central Asia or combined with South Asia.

The Organization for Economic Cooperation and Development (OECD): United States, Canada, European Union (Belgium, Luxembourg, Denmark, France, Germany, Greece, Ireland, Italy, Netherlands, Portugal, Spain, United Kingdom), Austria, Finland, Iceland, Malta, Norway, Sweden, Switzerland, Turkey, Australia, New Zealand, Japan.

Other developed countries: Israel and sometimes South Africa.

Measures

1 billion = 1,000 million	kg = kilogram(s)
Dollars ($) are US dollars	ha = hectare(s)
Calories = kilocalories	km = kilometre(s)
Tons = metric tons, i.e. 1,000 kilograms	ppb = parts per billion

Grain means cereals, e.g. wheat, barley, rice, maize, oats, sorghums, millets and other coarse grains. (It does not include grain legumes or pulses.)

Institutional Acronyms

CGIAR   Consultative Group on International Agricultural Research, at the World Bank, Washington DC (USA)

CIAT   Centro Internacional de Agricultura Tropical (Colombia)

CIFOR   Centre for International Forestry Research (Indonesia)

CIMMYT   Centro Internacional de Mejoramiento de Maíz y Trigo (Mexico)

CIP   Centro Internacional de la Papa (Peru)

FAO   Food and Agricultural Organization, Rome (Italy)

IARCs   International Agricultural Research Centres (funded by the CGIAR – see Appendix)

ICARDA   International Center for Agricultural Research in the Dry Areas (Syria)

ICLARM   International Center for Living Aquatic Resources Management (Philippines)

ICRAF   International Centre for Research in Agroforestry (Kenya)

ICRISAT   International Crops Research Institute for the Semi-Arid Tropics (India)

IDS   Institute of Development Studies, University of Sussex (UK)

IFPRI   International Food Policy Research Institute, Washington DC (USA)

IIED   International Institute for Environment and Development, London (UK)

IIMI   International Irrigation Management Institute (Sri Lanka)

IITA   International Institute of Tropical Agriculture (Nigeria)

ILRI   International Livestock Research Institute (Kenya)

IMF   International Monetary Fund, Washington DC (USA)

IPGRI   International Plant Genetic Resources Institute (Italy)

IRRI   International Rice Research Institute (Philippines)

ISNAR   International Service for National Agricultural Research (Netherlands)

LIFDCs   Low-income food-deficit countries

NGOs   Non-governmental organizations

NICs   Newly industrialized countries

UNEP   United Nations Environment Programme, Nairobi (Kenya)

UNHCR   United Nations High Commissioner for Refugees

UNICEF   United Nations Children's Fund, New York (USA)

UNRISD   United Nations Research Institute for Social Development

USAID   United States Agency for International Development, Washington DC (USA)

USDA   United States Department of Agriculture, Washington DC (USA)

WARDA   West Africa Rice Development Association (Côte d'Ivoire)

WHO   World Health Organization, Geneva (Switzerland)

Sources of Figures

Unless otherwise indicated, the figures are derived from data published by the Food and Agriculture Organization, Rome, in the form of two data diskettes:

1. FAOSTAT TS: SOFA '95, available with: FAO, *The State of Food and Agriculture, 1995*, Rome (Italy), Food and Agriculture Organization, 1995

2. FAOSTAT TS. AGROSTAT3. 1995

1 Hunger and Poverty

To die of hunger is the bitterest of fates
— Homer, the *Odyssey* [1]

Homer in his epic poem, the *Odyssey*, recounts how Odysseus and his companions have resisted the lure of the Sirens, sailed safely between Scylla and Charybdis, and have come to the island of Thrinacie, where the 'Sun-god's cattle and plump sheep graze'. Odysseus has been warned they are not to be harmed, but his companions succumb to the temptation. 'To die of hunger', declares Eurylochus, 'is the bitterest of fates.' They kill the cattle and feast. No sooner have they set sail again than Zeus sends a hurricane as a punishment. All perish except for Odysseus.

Today, there are more than three-quarters of a billion people who, like Odysseus's companions, live in a world where food is plentiful yet it is denied to them. If we were to add up all the world's production of food and then divide it equally among the world's population, each man, woman and child would receive a daily average of over 2,700 calories of energy [2]. Would this be enough? There are regional differences in energy requirements; people in cold climates need more calories to keep warm. Individuals also differ in their needs. Adults need more than children. Heavy manual labour requires about 4,000 calories a day, but 2,700 is enough for light activity [3]. Taking these considerations into account, a global average of 2,700 calories per day is enough to prevent hunger and probably sufficient for everyone to lead active, healthy lives.

Yet the harsh reality is great inequality. While in Western Europe and North America average supplies exceed 3,500 calories a day, they are less than two-thirds this amount in Sub-Saharan Africa and South Asia (Figure 1.1). Thirty-five developing countries, including nearly half the countries of Africa, have average supplies of less than 2,200 calories per day. According to recent estimates, over 800 million people, equivalent to 15 per cent of the world's population, get less than 2,000 calories per day and live a life of permanent or intermittent hunger and are chronically undernourished [4].

Unlike Odysseus's companions, many of the hungry are women and children. More than 180 million children under five years of age are underweight, that is they are well below the standard weight for their age. This represents a third of the under-fives in the developing countries. Young children crucially need food because they are growing fast and, once weaned, are liable to succumb to infections. And women, in addition to their own needs, require an adequate diet if they are to give birth to, and raise, healthy children.

Chronically undernourished people are short of the calories needed for their daily energy needs. Malnutrition results from lack of proteins, vitamins, minerals and other micro-nutrients in the diet. The most important sources of calories are cereals, and cereal diets that supply sufficient calories usually provide enough protein as well [5]. This is partly the reason why much of the data in this book refers to cereal production and consumption. But vegetables, fruit, livestock and fishery products are also important sources of proteins, vitamins and minerals. It is total food supply, measured in terms of quality as well as quantity, which determines whether people are adequately fed.

For many, undernutrition and malnutrition lead to death, not necessarily

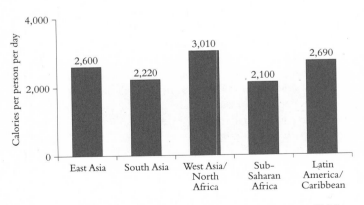

Figure 1.1 Average per capita calorie supplies in the developing world [6]

2

through starvation, although this may happen in famine situations, but because a poor diet reduces the capacity to fight disease [7]. Diarrhoea, measles, respiratory infections and malaria are common in many parts of the developing world. Well-fed people can fight them off; the malnourished, especially children, will succumb. Seventeen million children under five die each year and malnourishment contributes to at least one-third of these deaths (Figure 1.2). About 40 million children suffer from vitamin A deficiency [8]. As has been long known, lack of this vitamin can cause eye damage. Half a million children become partially or totally blind each year, and many subsequently die. And as recent research has shown, lack of vitamin A has an even more serious and pervasive effect, apparently reducing the ability of children's immune systems to cope with infection. Trials conducted in several developing countries in which children under five were given vitamin A supplementation reduced mortality from measles and diarrhoea by 25 per cent [9]. Lack of minerals in the diet can have equally severe effects. Iron deficiency is common in the developing world, affecting as many as a billion people. Anaemia caused by iron deficiency afflicts over 400 million women of childbearing age (fifteen to forty-nine years old). Anaemic women tend to produce stillborn or underweight children and are more likely to die in childbirth.

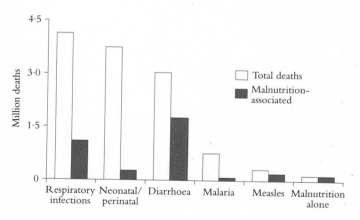

Figure 1.2 Causes of death among children under five years old in developing countries [10]

Hunger is particularly common in South Asia and Sub-Saharan Africa (Figure 1.3). Although the proportion has fallen in recent years, the number of undernourished and malnourished has risen as the population has grown. There were 20 million more underweight children at the end of the 1980s than at the beginning of the decade.

Paradoxically, hunger is common despite twenty years of rapidly declining world food prices (Figure 1.4). In many developing countries there is enough food to meet demand, yet large numbers of people still go hungry. Although food prices are low, they remain high relative to the earning capacity of the poor. Market demand is satisfied, but there are many who are unable to purchase the food they need and, hence, to whom the market is oblivious. India is reported 'self-sufficient' in food supply. Over 30 million tons of grain are surplus to the market and held as stocks. India exported 2 million tons of rice in 1995/6. Mr Balram Jakhar, India's agriculture minister, is quoted as saying: 'We have ensured abundant availability to domestic consumers, and now our goal is to boost foreign exchange earnings and ensure more remunerative prices for farmers' [11]. Yet more than 300 million Indians live below the poverty line and over half of these are chronically undernourished.

As the economist Amartya Sen points out, hunger occurs because, in one way or another, people are not entitled to the wherewithal to obtain food [12]. They may be unable to:

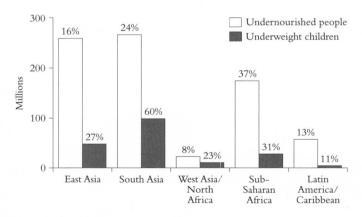

Figure 1.3 Hunger by region in the developing world [13]

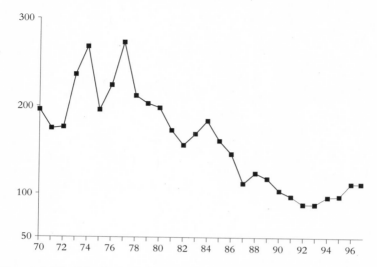

Figure 1.4 World food price index (in constant dollars) [14]

– Grow enough food on the land they own or rent or are otherwise entitled to cultivate;

– Buy enough food because their income is too low, or they are unable to borrow, beg or steal enough money;

or

– Acquire enough food as a gift or loan from relatives or neighbours, or through entitlement to government rations or aid donations.

Not surprisingly, hunger is closely related to poverty. Poor people have few or no assets, are unemployed or earn less than a living wage and thus cannot produce or buy the food they need. According to World Bank estimates, over 1·3 billion people, that is, one-third of the developing world's population, were in poverty in 1993, defined as living at less than $1·00 a day. Most of South Asia's poor are in India, where they comprise over 50 per cent of the population, and nearly half of Sub-Saharan Africa's population are also in poverty. Most of East Asia's poor live in China, but they amount to less than 20 per cent of the population (Figure 1.5).

To the casual observer, poverty seems to be worst in the cities but, in

5

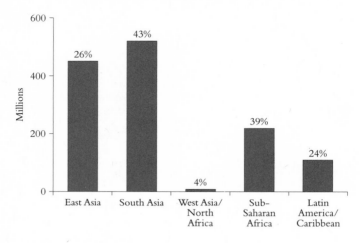

Figure 1.5 Poverty in the developing world in 1993 [15]

reality, the urban poor fare better. Although the cost of living may be low in rural areas, there are fewer opportunities to make a living. At the extreme, the urban poor can at least beg or steal. To quote one statistic, the incidence of malnutrition is five times higher in the Sierra of Peru than in the capital, Lima. About 130 million of the poorest 20 per cent of developing-country populations live in urban settlements, most of them in slums and squatter settlements (Figure 1.6). Yet 650 million of the poorest live in rural areas. In Sub-Saharan Africa and Asia most of the poor are rural poor: in Kenya and Indonesia the proportion is over 90 per cent [16]. Some live in rural areas with high agricultural potential and high population densities – the Gangetic plain of India and the island of Java. But the majority, about 370 million, live where the agricultural potential is low and natural resources are poor, as in the Andean highlands and the Sahel.

One of the biggest concentrations of poverty is in northern India. In three states – Bihar, Madhya Pradesh, and Uttar Pradesh – about half the rural population lives beneath the poverty line, only 10 per cent of women are literate and rural infant mortality rates are, in Uttar Pradesh, as high as 150 deaths per 1,000 live births [17]. Most of the rural poor have little or no land [18]. In Bangladesh more than 70 per cent of households owning

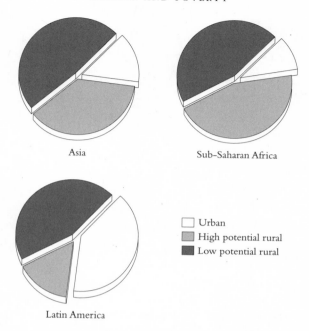

Asia

Sub-Saharan Africa

☐ Urban
▨ High potential rural
▦ Low potential rural

Latin America

Figure 1.6 The locations of the poorest people [19]

less than a hectare are poor and this rises to 93 per cent for the landless (Figure 1.7). When they do own land it is often unproductive and is rarely irrigated. They find themselves unable to improve their land because of lack of income and access to credit. Despite using up to 80 per cent of their income to obtain food and a similar proportion of working time for its production and preparation, they remain undernourished [20].

Poverty is common among ethnic and other minorities, for example among the indigenous people of Latin America and the scheduled castes and tribal peoples of India. A large number of the poorest and most oppressed are women. They often shoulder a disproportionate share of the workload: in Nepal while poor men work, on average, seven and a half hours a day, poor women work eleven [21]. Their schooling is inadequate (for the developing countries as a whole there are 86 females for every 100 males in primary education and 75 in secondary [22], and they have few opportunities to improve their income or status. Boys are

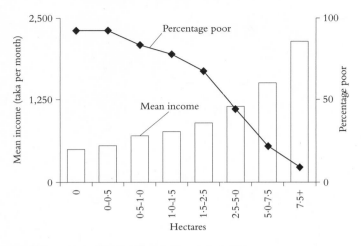

Figure 1.7 Poverty and landholding in Bangladesh [23]

favoured over girls in India. Abortion of female foetuses, female infanticide and poorer diets for girl children result in an overall sex ratio of 992 women for every 1,000 men, falling to as low as 882 women in states such as Uttar Pradesh [24]. Over 70 per cent of the poor are women and some of the poorest households are headed by women: they are burdened with a high proportion of dependants and typically have been left with poor resources although, as experience shows, provision of credit and some training can produce significant increases in income, especially through micro-enterprise and non-farm employment (see Chapter 15) [25].

These statistics provide an overview of the reality of millions of people's lives, but a full picture of the deprivation and misery they suffer can only be obtained from close acquaintance. Dominique Lapierre in his well-known book *City of Joy* describes, in the opening chapter, the circum-stances which drive a West Bengal family to migrate to the slums of Calcutta [26]. Thirty-two-year-old Hasari Pal lives with his wife, Aloka, and three children, his parents, two younger brothers and their families – in all, sixteen people. His father Prodip had once owned over three hectares of good riceland but a large landowner, by bribing the judge in a court action, had wrested the land from him. The family was reduced to owning less than half a hectare, sharecropping another plot and growing vegetables, tending a few fruit trees and caring for a buffalo, two cows and two goats on the land around their house. They survived until, one year, pests destroyed the rice crop. The land was mortgaged to the village moneylender and one brother became an agricultural labourer. For two more years they struggled on, and then a storm brought down all the mangoes and coconuts. They had to sell the buffalo and one of the cows. Hasari's youngest brother began to cough blood and money was needed for the doctor's fees and medicine. In the same year, his youngest sister was married. The dowry and the ceremony cost the family 2,000 rupees (about $200). They spent their savings and pawned the family jewellery. Then came the final blow: the monsoon failed, the young rice crop withered in the field, the village wells dried up. The household had no more assets, nothing they could mortgage or pawn, and so Hasari Pal, Aloka and their children set off, with trepidation, to find a new life in Calcutta.

Lapierre's account is fictional, but based on careful observation. In his book *Jorimon*, Muhammad Yunus, the founder of the Grameen Bank, presents real life stories of poor women in rural Bangladesh [27]. Koituri's story is not uncommon. She was married at thirteen to Joynal, a twenty-year-old who worked as a labourer. Joynal turned out to be a cruel husband: he beat his wife, constantly demanding a dowry although this had not been part of the marriage negotiations. Although she provided two sons, the beatings continued and eventually she could stand it no longer. She moved to her father's compound, and worked in the houses of better-off neighbours, husking paddy, cleaning the cowsheds and doing domestic chores. For this she received two meals a day and a kilo of rice. Koituri, unlike the Pals, did not migrate to the city. In Chapter 15 I describe how a series of loans from the Grameen Bank helped her to achieve independence and security.

The first question we ought to ask ourselves is: why should we be concerned? Probably everyone who reads this book is getting an adequate diet. Does it matter to us that others are not so fortunate? Does it matter to the industrialized countries that many people in the developing countries are malnourished? Part of the answer to these questions is political. The end of the Cold War has not brought about an increase in global stability. While conflict between East and West has declined, there is a fast-growing divide between the world of the peoples, countries and regions who 'belong' in global power terms and those who are excluded. Over 2 billion people in the world regularly watch television. For the rich, the images on their screens provide a constant reminder of the horrors of natural disasters, civil war and famine. For the poor, the screens portray the everyday luxuries of the affluent and well fed. Globally, the consequence is a potentially explosive mix of fears, threats and unsatisfied hopes.

Yet this growing conflict receives relatively little attention in the industrialized countries. The end of the Cold War has made political agendas inward-looking. Governments struggling with rising costs for welfare payments are paying little attention to the needs of poor nations overseas. The volume of agricultural aid going to developing countries is declining in real terms (Figure 1.8). And the industrial world's attention to external problems is being focused on the former eastern-block countries.

Reductions in aid may be justifiable in the short term but are not in the long-term or even medium-term interest of the industrialized countries. An increasingly polarized world will result in growing political unrest. Already the consequences of economic stagnation, population growth, environmental degradation and civil war are producing unprecedented movements of peoples. There are currently some 14 million refugees in

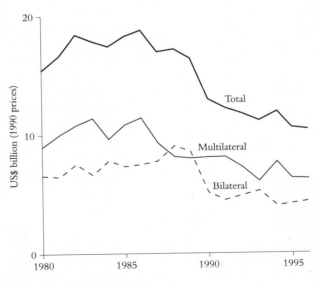

Figure 1.8 World agricultural aid [28]

10

need of assistance living in foreign countries and at least double that number who are refugees or displaced persons within their own countries [29]. Unless the developing countries are helped to realize sufficient food, employment and shelter for their growing populations or to gain the means to purchase the food internationally, the political stability of the world will be further undermined. In today's world, poverty and hunger, however remote, affect us all.

At the same time, the growing interconnectedness of the world – the process commonly referred to as globalization – holds the promise of alleviating, if not eliminating, poverty and hunger. Modern technology and the accompanying research and development are now disseminated worldwide through international research networks. Much is in the hands of multinational companies, yet many powerful, advanced technologies are often small in scale, readily transferable and becoming inexpensive with mass use.

Capital, too, is commonly invested on a global scale. Multinational finance corporations are growing in power as capital becomes more mobile and financial markets are deregulated. There is little incentive for such corporations to invest in countries perceived as poorly endowed with human and physical resources. Nevertheless, the worldwide availability of capital provides an enhanced potential for new economic opportunities. New systems of local banking, pioneered by the Grameen Bank of Bangladesh, which rely on self-regulating groups of savers are increasingly able to provide the small loans that the poor require.

Alongside capital, trade operates through a great variety of global markets. While trading blocks such as the European Union continue to prevent easy access of goods from developing countries, the outcome of the Uruguay Round of negotiations is likely to create new opportunities for manufactured goods and agricultural exports (see Chapter 15). And finally governance, the form and manner of government, is also becoming globalized. Traditional patterns are becoming eroded. The supremacy of national governments is being challenged from within by ethnic and religious groupings and from outside by supranational institutions, such as the International Monetary Fund (IMF), the World Bank, the World Trade Organization and the European Union. At the same time, non-governmental organizations (NGOs) are beginning to operate globally, pressing for citizens' rights, more development and the elimination of poverty.

Thus globalization, while threatening, on the one hand, to concentrate power and increase division, on the other contains the economic and technological potential to transform the lives of rich and poor alike. I believe the elimination of poverty and hunger is a goal within our capacity. Much depends on where our priorities lie and, in particular, whether there is sufficient access by the poor to the economic opportunities created by the products of the new technologies. In this book I set out an approach and an agenda aimed at satisfying the aspirations of the poor by bringing the power of modern technology to bear on the problem of providing food security for all in the twenty-first century.

Notes

[1] Translated in H. B. Ash, 1941, *Lucius Junius Moderatus Columella on Agriculture*, Vol. 1 (12,342), Cambridge, Mass., Harvard University Press (Loeb Classical Library), London, Heinemann

[2] N. Alexandratos (ed.), 1995, *World Agriculture: Towards 2010. An FAO study*, Chichester (UK), Wiley & Sons

[3] FAO/WHO, 1973, *Energy and Protein Requirements; Report of a Joint FAO/WHO Ad Hoc Expert Committee*, Geneva (Switzerland)

[4] FAO, 1992, *World Food Supplies and Prevalence of Chronic Undernutrition in Developing Regions as Assessed in 1992*, Rome, Food and Agriculture Organization (Document ESS/MISC/1992); FAO, 1996, *Sixth World Food Survey*, Rome (Italy), Food and Agriculture Organization

[5] D. Grigg, 1993, *The World Food Problem* (2nd edn), Oxford, Blackwell; T. Dyson, 1996, *Population and Food: Global Trends and Future Prospects*, London, Routledge

[6] The figures omit food fed to livestock. United Nations Administrative Committee on Coordination/Sub-Committee on Nutrition, 1992, *Second Report on the World Nutrition Situation*, Vol. 1, *Global and Regional Results*, Geneva (Switzerland); Food and Agriculture Organization, 1992, op. cit.; Alexandratos, op. cit.

[7] World Resources Institute, 1992, *World Resources 1992– 1993*, Oxford, Oxford University Press

[8] United Nations Administrative Committee on Coordination/Sub-Committee on Nutrition, 1992, op. cit.; Unicef, 1990, *Strategy for Improved Nutrition of Children and Women in Developing Countries*, New York, NY, United Nations Children's Fund

[9] G. Beaton, R. Martorell, K. J. Aronson, B. Edmonston, G. McCabe, A. C. Ross and B. Harvey, 1993, *Effectiveness of Vitamin A Supplementation in the Control of Young Child Morbidity and Mortality in Developing Countries*, Geneva (Switzerland) (Nutrition Policy Discussion Paper No. 13, State of the Art Series, ACC/SCN)

[10] WHO, 1995, *The World Health Report 1995: Bridging the Gap*, Geneva (Switzerland), World Health Organization

[11] *Financial Times*, 1995, October 14/15

[12] A. Sen, 1981, *Poverty and Famines: An Essay on Entitlement and Deprivation*, Oxford, Clarendon Press

[13] Underweight children (under five years old) are less than two standard deviations below the standard weight for their height. The undernourished are those individuals whose average daily intake of calories is not sufficient to 'maintain body weight and support light activity' (set at 1·54 times the basal metabolic rate). The threshold varies, because of differences, from 1,760 cal/day in Asia to 1,985 cal/day in Latin America. Sources: United Nations Administrative Committee on Coordination/Sub-Committee on Nutrition, 1992, op. cit.; FAO, 1992, op. cit.; Grigg, op. cit.

[14] World Bank, 1998, *World Development Indicators 1998*, Washington, DC, World Bank

[15] M. Ravallion and S. Chen, 'What can new survey data tell us about recent changes in distribution and poverty', *World Bank Economic Review*, 11, 357–382

[16] ibid.

[17] R. Chambers, N. C. Saxena and T. Shah, 1989, *To the Hands of the Poor: Water and Trees*, New Delhi (India), Oxford and IBH Publishing Co.

[18] R. Sinha, 1984, *Landlessness: A Growing Problem*, Rome (Italy), Food and Agriculture Organization

[19] The poorest 20 per cent of the population. H. J. Leonard, 1989, 'Overview: environment and the poor', in H. J. Leonard, *Environment and the Poor: Development Strategies for a Common Agenda*, Washington DC, Overseas Development Council (Third World Policy Perspectives No. 11), compiled from the World Bank, 1988, *Social Indicators of Development*, Baltimore, Md, Johns Hopkins University Press, and unpublished data from the World Bank and the International Food Policy Research Institute, Washington DC

[20] I. Jazairy, M. Alamgir and T. Panuccio, 1992, *The State of the World Rural Poverty – An Inquiry into Its Causes and Consequences*, New York, NY, New York University Press; M. Lipton and R. Longhurst, 1989, *New Seeds and Poor People*, London, Unwin Hyman

[21] Ravallion and Chen, op. cit.

[22] Unesco, 1991, *World Education Report 1991*, Paris (France), United Nations Educational, Scientific and Cultural Organization

[23] M. Ravallion, 1989, 'Land-contingent poverty alleviation schemes', *World Development*, 17, 1223–33

[24] A. N. Agrawal, H. O. Varma and R. C. Gupta, 1993, *India: Economic Information Yearbook 1992–1993* (7th edn), New Delhi (India), National Publishing House

[25] FAO, 1996, *Women Feed the World*, Rome, Food and Agriculture Organization (World Food Summit briefing note)

[26] D. Lapierre, 1986, *City of Joy*, London, Arrow Books

[27] M. Yunus (ed.), 1984, *Jorimon of Beltoil Village and Others: In Search of a Future*, Dhaka (Bangladesh), Grameen Bank

[28] FAO 1996, *Investing in Food Security*, Rome, Food and Agriculture Organization (Briefing paper for the World Food Summit); FAO 1998, Rome, Food and Agriculture Organization (The Statistics Division)

[29] UNHCR, 1995, *State of the World's Refugees: In Search of Solutions*, Oxford, Oxford University Press

2 The Year 2020

For most of humanity, the world will not be a pleasant place in the twenty-first century. Yet it does not have to be this way. With foresight and decisive action, a better world can be created for all people.

> – The International Food Policy Research Institute,
> *A 2020 Vision For Food, Agriculture, and the Environment*

If nothing new is done, the numbers of poor and hungry will grow. Partly this is because most populations in the developing world are still rapidly growing. I am not a Malthusian, or even a neo-Malthusian, and I do not intend to enter the current debate on the relevance of Malthus's ideas to today's problems [1]. But the simple fact is that by the year 2020, less than twenty-five years from now, there will be about an extra 2·5 billion people in the developing world who will require food. This is additional to the three-quarters of a billion people who are chronically undernourished today.

While the growth rate of the world's population has declined from a high of about 2·0 per cent a year during the late 1960s to 1·6 per cent in the 1990s, the size of the current annual increment is unprecedented [2]. Nearly 900 million people will be added to the world's population in the 1990s, the largest increase for any decade in history [3]. Until well into the next century, a further 80 million people will be added each year, close to a quarter of a million people per day (Figure 2.1). If the proportion of the population of the developing countries deprived of an adequate diet remains the same, the number undernourished by the year 2020 could be greater than 1·4 billion.

The United Nations and the World Bank have published a range of scenarios for the growth of the world population. These are based on different estimates of the speed with which birth rates will fall. The scenarios produce widely divergent forecasts for the population at the end of the next century – nearly 20 billion at one extreme; 6 billion at the other.

15

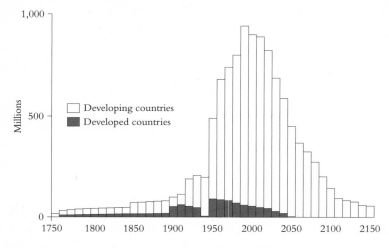

Figure 2.1 Increments by decade to the world's population, 1750–2150 [4]

One of the factors which determines the fall in birth rate is agricultural production, but the relationship is complex and subject to much debate. Esther Boserup in a famous book, *The Conditions of Agricultural Growth*, argues that population growth can stimulate agricultural innovation [5]. Her case is borne out by many historical and contemporary examples. In Kakamega and other districts in the Western Province of Kenya, rapid population growth has produced densities of the order of 700 persons per km², yet the land, far from being degraded, is highly productive. When Robert Chambers, of the Institute of Development Studies, and I visited the region in 1988 we saw tiny hillside plots, intensively cultivated with a great variety of trees and annual food crops – on one quarter-of-a-hectare plot I counted some thirty different useful species of plant, plus a cow and a calf (Figure 2.2). Innovation here is partly driven by population pressure. However, as Tim Dyson of the London School of Economics cautions us, the sheer speed of contemporary population growth – populations doubling in less than twenty years in some parts of Sub-Saharan Africa – places severe limits on what can be achieved by farmers' efforts unaided [6]. Much will depend on the availability of appropriate technological assistance and on supportive policies. In Kakamega the farmers have secure ownership of their land.

There is also good evidence that intensive agricultural production can lead to falls in the birth rate. One of the fastest declines in population growth occurred in the Chiang Mai valley in northern Thailand during the 1970s and 1980s. From a high of over 2·5 per cent in the 1960s the population is now growing at less than 1 per cent [7]. Part of the reason was a vigorous government policy of providing readily available contraception. But, as family planning experts are only too aware, take-up of contraception depends on motivation. In this case the desire to limit family size arose, indirectly, from a rapid intensification of agriculture in the valley, which in turn was stimulated by increased availability of irrigation. Farmers began to plant two or even three crops a year, following the traditional rice crop with high-value, although labour-intensive, crops of vegetables. Women soon became aware of the relative returns from working on vegetable cultivation and the losses of income they incurred during pregnancy and child-raising. They began to have fewer offspring and to invest their earnings in their children's future.

Birth rates in many parts of the world are falling faster than was earlier expected. The decline began almost simultaneously in Asia, Latin America and West Asia/North Africa around 1970 and has been fastest in Asia. China has pursued a highly interventionist and aggressive policy, encouraging the adoption of birth-control methods and, since 1979, penalizing those who have more than one child per family. But there have been similarly impressive declines in more liberal environments. In Bangladesh fertility has fallen from 7·0 births per woman in the early 1970s to 4·7 in the 1990s and in India it is now down to 3·9 [8]. Even in parts of Sub-Saharan Africa there has been a significant decline: in Kenya from 8·1 in the 1970s to 6·3 now, although there is a long way to go before the rates equate with those of Asia.

The three United Nations projections for 2020 are not very divergent: plus or minus 450 million people about the medium projection, which predicts a global population of about 8 billion, of whom some 6·7 billion will be in the developing countries. By then the world's population growth rate will have slowed to 1·0 per cent and the annual increment will be falling. If some experts' optimism about falling birth rates is confirmed, the developing-country total could be close to the bottom of the range, at 6·2 billion, but the medium projection is probably the safest to assume for planning purposes for the next thirty years [9].

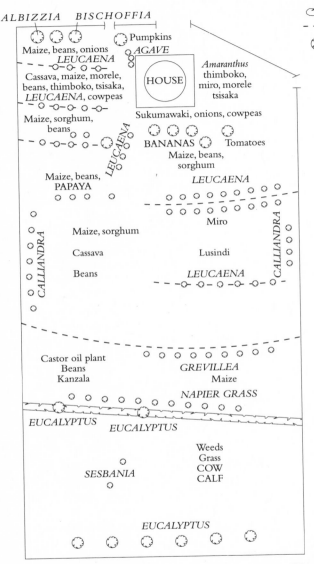

Figure 2.2 A modern quarter-hectare farm, Kakamega, Western provinces, Kenya [10]

Despite the significant falls in Asian fertility in recent years, the sheer size of the Asian population will generate very large increments for some time to come. By 2020 the population of South Asia will have grown to about 2 billion while China's population will stand at 1·5 billion, equivalent to that of India. However, the fastest growth rate will be in Sub-Saharan Africa, primarily because fertility levels, except in a few countries (Kenya, Botswana, Zimbabwe and South Africa), have yet to fall. By 2020 the population of Sub-Saharan Africa will have more than doubled (Figure 2.3). An inescapable conclusion from these population projections is that the biggest increases will be in those regions of the world where the poor and hungry are currently concentrated.

A question often asked is whether these projections will be significantly altered by the AIDS epidemic which is affecting large populations in parts of Asia and Sub-Saharan Africa. About 12 million people in the developing countries are currently infected with the AIDS virus HIV. In an attempt to answer this question the United Nations has examined the possible impacts in fifteen African countries where the prevalence of HIV is greater than 1 per cent of the population [11]. As expected, these countries will experience a lowered average life expectancy and a higher death rate and, as a consequence, a significantly lower growth rate. The UN predicts that

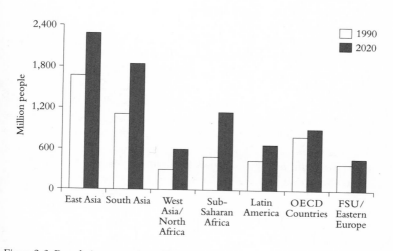

Figure 2.3 Population growth to 2020 [12]

by 2005 their total population will be some 12 million less than would have occurred in the absence of AIDS. This will result in a lower food demand; but there are other considerations to bear in mind. In the immediate future, the worst-hit countries will lose a high proportion of their young and able-bodied population, those most able to contribute to future food production and to growth in the economy [13].

By the year 2020 not only will there be more people, a much lower proportion will be living in rural areas. Over the past four decades there has been a dramatic growth in urbanization. Gigantic urban complexes have been created in the developing countries. São Paulo now has a population of over 16 million, Mexico City over 15 million, Bombay and Shanghai over 14 million, Jakarta and Calcutta over 11 million [14]. Urban populations have grown because, contrary to the casual impressions of the Western visitor, living conditions, on average, have been better than in the countryside. There are more opportunities for making a living, through employment in manufacturing, construction, trade and services, or by begging or stealing. Food tends to be cheaper and medical facilities more readily available (Figure 2.4). In *City of Joy*, when Hasari Pal and his family arrive in Calcutta, they find a place to camp on the pavement. Hasari gives blood for a few rupees and purchases some bananas, enough to keep his family fed for a few days. Then, by chance, he meets up with someone from his home village, a rickshaw puller who persuades the owner of the rickshaws to let Hasari work in the place of a puller who has just died. It is a hard life, and Hasari eventually dies from the strain of the work, but he has saved enough for a dowry for his daughter and his family survives. The opportunity is there, for some at least, to slowly progress from the pavement to the slum and to the beginnings of a decent livelihood.

The higher child survival and longer life expectancy, coupled with massive in-migration from the countryside, are causing urban populations to grow more rapidly than rural populations. In the 1950s over 80 per cent of the population of the developing countries were living in rural areas; by 1985 this had shrunk to just over 70 per cent; by 2020 it will be less than 45 per cent. Over 3·5 billion people in the developing world will be urban dwellers in 2020 and most will be food consumers rather than producers. They will require cheap food.

What is the prognosis for feeding the world's population in the twenty-first century? It is not possible to foresee the situation in the latter half of the next century; that would be an exercise in crystal-ball gazing. Predicting

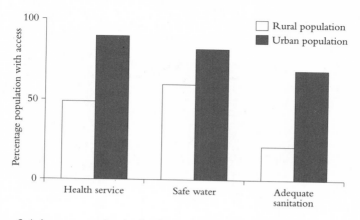

Figure 2.4 Access to services in developing countries [15]

the next twenty-three years is more feasible, and this will be the most critical period; after 2020 the annual increments in world population will begin to decrease significantly. If we can achieve a well-fed world by then it should be possible to meet future demands, provided the resource base has been adequately protected.

Producing forecasts of world food production is complicated. There are numerous variables – and many unknowns – that have to be considered, and they interrelate with one another in complex and often circuitous ways. There are, in systems jargon, many feedback-loops which can bring about stability, growth or collapse. For example, if population size stimulates agricultural and hence food production the net effect is to stabilize, or allow for the moderate growth of, food availability per capita. On the other hand, if population size results in increased soil erosion it will produce rapidly declining food per capita. Both effects are moderated, however, if food availability per capita produces a decline in birth rates (Figure 2.5). Unfortunately, most current models do not attempt this level of sophistication.

At the time of writing, most attention is being paid to three models. These differ in their scope and level of detail [16]. The United Nations Food and Agriculture Organization (FAO) study [17] models virtually every country and covers a wide range of food products; the International Food Policy Research Institute (IFPRI) study [18] focuses on thirty-five

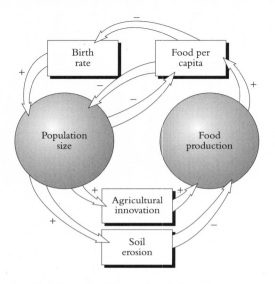

Figure 2.5 Feedback loops indicating possible linkages between population size and food production [19]

countries or regions and on staple crops and livestock; while the World Bank World Grains Model [20] further restricts itself to wheat, rice and coarse grains as a group. What the models all have in common is their attempt to mimic the workings of a world market in which demand for food is met by supply.

As with all models their outcomes depend on the assumption built into them and the worth of the baseline data. They are essentially econometric models: they assume that supply meets demand in a market economy and they are constructed to ensure this occurs:

– Regional or country food demand is calculated by multiplying population size by per capita demand. Population estimates are based on the UN medium variant and per capita demand is a function of per capita income, which, in turn, is derived with some modifications from the World Bank projections of gross domestic product (GDP).
– Food production is determined by multiplying the amount of cropped

land by the yield, which in the IFPRI model is predicted to increase annually by approximately 1·5−1·8 per cent per year, depending on the cereal.

− For each country (or region) and each year supply is equated to demand through market prices.

In each of the three models the forecast is reasonably optimistic. The world population growth rate is matched by a similar growth in food production. World food prices continue to decline. However, the developing countries as a whole will not be able to meet their market demand. In the IFPRI model the total shortfall is some 190 million tons and the model predicts this can be met by imports from the developed countries (Figure 2.6).

The developing countries imported about 3 per cent of the total grain they consumed in the 1960s and 1970s. This increased to 9 per cent or 91 million tons in 1990. The predicted imports for 2020 are double this figure and comprise 11 per cent of consumption. Most of the increase in imports

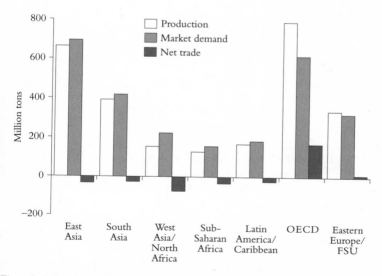

Figure 2.6 Grain production and market demand (for human and livestock feed) from the IFPRI model for the year 2020 [21]

will be of wheat, maize and other coarse grains; the market demand for rice will be largely satisfied by domestic production. For the developed countries the increase in net exports represents an 11 per cent growth in cereal production.

The relative optimism in this and the other models is based on several assumptions. The first is a lower rate of growth, or even a fall, in the global average per capita consumption of grain. This is partly for arithmetic reasons: relatively more people will be living in the poorer developing countries. By 2020 nearly 40 per cent of the world's population will live in South Asia and Sub-Saharan Africa. It is also partly a result of rising incomes in the better-off developing countries. Economies typically experience several phases in the growth of grain consumption. As incomes begin to rise there is a rapid increase in consumption but then, as incomes rise still further, growth slows because consumers turn to other foodstuffs. In Japan, for example, grain consumption grew by over 3 per cent per year in the 1960s, but fell to about 2·5 per cent in the 1970s and to only 0·6 per cent in the 1980s [22]. While rapid increases are still occurring in countries such as India and Pakistan, a pattern of declining growth is now emerging in a number of developing countries, notably China, Brazil, Indonesia and North Africa. In China GDP grew substantially faster in the 1980s than in the 1970s but growth in grain consumption fell from just over 5 per cent to little more than 2 per cent per year [23].

With rising incomes the use of grain also changes: an increasing pro- portion goes to livestock feed. Over 60 per cent of cereal production in the developed countries is consumed by livestock, compared with only 17 per cent in the developing countries [24]. Most of the modest future increase in grain consumption in the developed countries is not for human food but for livestock feed, industrial and other uses. This trend in the developing countries is most marked in the rapidly growing economies of East Asia; in South Korea and in Malaysia 40 per cent of grain consumption now goes to livestock feed. Grain consumption in the form of human food will grow at nearly 2 per cent in the developing countries as a whole, but at nearly 4 per cent for livestock feed [25]. This is roughly the same as the projected growth in livestock production, which will be predominantly in the form of pigs and poultry, both heavily dependent on grain for their feed.

Overall, the average per capita grain consumption in the developing countries has not increased since 1984 and is expected to rise by only about 0·4 per cent per year in future. Average world consumption is also likely to be affected by the situation in Eastern Europe and the countries of the former Soviet Union, where the current levels of grain consumption are very high compared with those of Western Europe, largely due to inefficiency and waste. With a growing free-market economy and declining subsidies, grain consumption will fall [26]. The net effect of these various factors is a slowing in the global average per capita grain consumption. This means that world market demand can be met, provided the increase in total grain production keeps pace with population growth.

Two more assumptions are critical to the optimistic scenarios of the models:

— First, that the developed countries will increase their food production to meet the needs of the developing countries; and
— Second, that the latter will be able to pay for increased imports.

According to the IFPRI model the required growth of developed-country grain production is only about 1 per cent per year. Partly this is due to the predicted decline in grain consumption in Eastern Europe and the former Soviet Union and increased production there, as privatization, smaller production units and importation of modern technology begin to have an effect. According to the model this region will no longer have need of imports (nearly 40 million tons in 1991 [27]) and will become a modest exporter — of about 15 million tons.

The second assumption is based on optimistic World Bank forecasts for income growth in the developing countries. In the 1980s income growth was relatively poor. Although there was high income growth in East Asia, there were significant declines in growth rates in Sub-Saharan Africa and the rest of the developing world. According to the World Bank this will change, as the current programmes of structural adjustments and economic reform take effect. The bank predicts significant income growth in the developing countries, including a low but positive growth rate in Sub-Saharan Africa (Figure 2.7).

Of the estimated 190 million tons of net grain imports to the developing countries in 2020, over 140 million tons will go to satisfy demand in East Asia, West Asia/North Africa and Latin America. This will be primarily wheat for human consumption and maize and other coarse grains for pig and poultry feed. Especially in East Asia a rising demand for wheat for

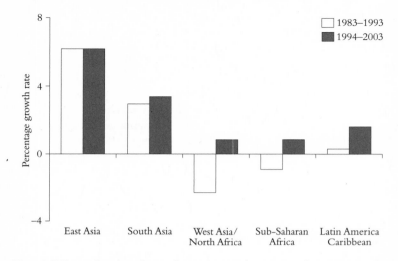

Figure 2.7 Past and projected growth rates in gross domestic product per person [28]

bread and pasta will be difficult to meet from within the region since much of the land is more suitable for rice production. Increasing incomes in these regions will result in dietary changes, in particular a growth of livestock consumption, as well as the capacity to pay for the necessary imports. However, in Sub-Saharan Africa and South Asia the proportion of grain going to livestock will remain very small and the imports will be required for human food.

Although overall the models are optimistic, there are pessimistic scenarios for significant regions of the developing world. Food production in Sub-Saharan Africa will be hard pressed to keep up with population increase for a long time to come. According to the IFPRI model, by the year 2020 the excess of market demand for grain over production will be nearly 26 million tons; this compares with current net imports of 9 million tons. And South Asia will require more than 22 million tons, compared with 1 million tons today.

Inevitably, models of this kind raise more question than they answer. The most important omission from the calculations is the food needs of the poor and hungry. As in the real world, they are simply priced out of the market and their needs are 'hidden'. The gap between demand and supply which the model closes is the market gap. If we convert the market availability predicted by the IFPRI model to calories per person per day, the regional averages are as in Figure 2.8. The improvement over current calories is slight. Then, as now, there is a substantial hidden food gap, particularly in Sub-Saharan Africa, where the average calorie availability remains below 2,200 calories per person per day, and in South Asia.

This hidden food gap is the cereal requirement to meet the energy need of the population, less the sum of domestic production and imports. Peter Hazell of IFPRI has calculated the total cereal requirement by assuming a minimum need per person of 3,000 cereal calories per day – which covers food, livestock feed, seed, storage losses and waste during processing. Using a conversion rate of one kilogram of cereals equivalent to 3,600 calories, each person then requires a little over 300 kg of cereals per year. The difference between the total need for the whole population and the market demand in the IFPRI model is then the 'hidden gap' (Table 2.1).

The assumption of 3,000 calories per person per day (or 300 kg of cereal per year) is hardly generous but it leads to total food gaps in terms of cereals of 214 million tons for Sub-Saharan Africa and 183 million tons

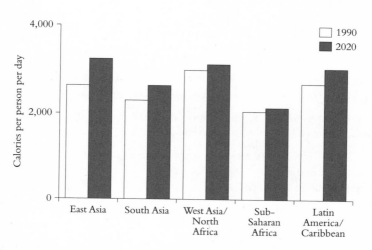

Figure 2.8 Predicted average per capita calorie supplies in the developing world in the year 2020 [29]

Table 2.1 The hidden and total food needs (in million tons) by the year 2020, assuming a cereal need equivalent to 3,000 calories per person per day [30]

Zone	Hidden food gap★	Imports	Total food gap
East Asia	–	55·8	55·8
South Asia	160·0	22·7	182·7
West Asia/North Africa	–	68·5	68·5
Sub-Saharan Africa	187·5	26·1	213·6
Latin America/Caribbean	11·6	15·0	26·6
Total developing countries	359·1	188·1	547·2

★ Hidden food gap: need less production and imports.

for South Asia in 2020. If all this food were to be supplied by the developed countries it would require nearly 550 million tons, nearly three times that predicted by the market model.

In human terms, the hidden gap can be translated into a persistence of large numbers of malnourished children. By 2020 the total numbers will have declined slightly from the current 180 million to 155 million, but in Sub-Saharan Africa they will have increased by nearly 50 per cent (Figure 2.9). And, probably, there will still be close to three-quarters of a billion people chronically undernourished (the FAO model predicts over 600 million) [31].

In conclusion, the models, while conveying a message which in some respects is optimistic, leave a number of crucial questions unanswered:

– Will GDP in the developing countries rise at the rate predicted, increasing demand and the ability to pay for imports?

– Will the developed countries be willing to pay the environmental and other costs of the increased production of cereals to meet the developing-country market demand?

– Will cereal yields in the developing countries increase as predicted?

and, most important of all,

– How are the 'hidden' needs of the poor of the developing countries to be met?

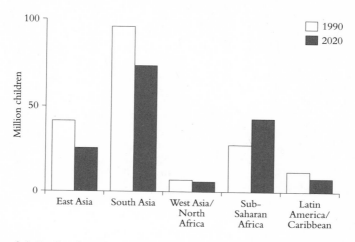

Figure 2.9 Predicted numbers of malnourished children from the IFPRI model [32]

Notes

[1] T. Dyson, 1996, *Population and Food: Global Trends and Future Prospects*, London, Routledge

[2] United Nations, 1995, *World Population Prospects. The 1994 Revision*, New York, NY, United Nations

[3] E. Bos, M. T. Vu, A. Levin and R. Bulatao, 1992, *World Population Projections*, 1992–1993 edn, Baltimore, Md, Johns Hopkins University Press, for the World Bank; United Nations, 1993, *World Population Prospects: The 1992 Revision*, New York, NY, United Nations

[4] J. Bongaarts, 1995, 'Global and regional population projects to 2025', in N. Islam (ed.), *Population and Food in the Early Twenty-first Century: Meeting Future Food Demands of an Increasing Population*, Washington DC, International Food Policy Research Institute, pp. 7–16

[5] E. Boserup, 1965, *The Conditions of Agricultural Growth*, London, Allen & Unwin

[6] Dyson, op. cit.

[7] P. Gypmantasiri, A. Wiboonpongse, B. Rerkasem, I. Craig, K. Rerkasem, L. Ganjanapan, M. Titayawan, M. Seetisarn, P. Thani, R. Jaisaard, S. Ongprasert, T. Radnachaless and G. R. Conway, 1980, *An Interdisciplinary Perspective of Cropping Systems in the Chiang Mai Valley: Key Questions for Research*, Chiang Mai (Thailand), Faculty of Agriculture, University of Chiang Mai

[8] B. Robey, S. Rustein and L. Morris, 1993, 'The fertility decline in developing countries', *Scientific American*, 269, 30–37

[9] P. Seckler and D. Cox, 1994, *Population Projections by the United Nations and the World Bank: Zero Growth in Forty Years*, Arlington, Va, Center for Economic Policy Studies, Winrock International Institute for Agricultural Development

[10] G. R. Conway and E. B. Barbier, 1990, *After the Green Revolution: Sustainable Agriculture for Development*, London, Earthscan

[11] United Nations, 1993, op. cit.

[12] United Nations, 1995, op. cit.

[13] K. A. Hamilton, 1993, 'The HIV and AIDS pandemic as a foreign policy concern', *Washington Quarterly*, 17, 201–15

[14] United Nations, 1995, *World Urbanization Prospects: The 1994 Revision*, New York, NY, United Nations

[15] UNDP, 1992, *Human Development Report 1992*, New York, NY, United Nations Development Program

[16] Queen Elizabeth House, 1996, *World Cereals Markets: A Review of the Main Models*, Oxford, Food Studies Group, Queen Elizabeth House (mimeo.)

[17] N. Alexandratos (ed.), 1995a, *World Agriculture: Towards 2010. An FAO Study*, Chichester (UK), Wiley & Sons

[18] M. W. Rosengrant, M. Agcaoili-Sombilla and N. D. Perez, 1995, *Global Food Projections to 2020: Implications for Investment*, Washington DC, International Food Policy Research Institute

[19] To interpret the diagram note the direction of the arrow and the sign: soil erosion negatively affects food production – the more the erosion the less the food production, and conversely the less the erosion the more the production; agricultural innovation, however, positively affects food production – the more the innovation the more the production, and conversely the less the innovation the less the production.

[20] D. O. Mitchell and M. D. Ingco, 1995, 'Global and regional food demand and supply prospects', in Islam, op. cit., pp. 49–60

[21] Rosengrant *et al.*, op. cit.

[22] Mitchell and Ingco, op. cit.

[23] ibid.

[24] Alexandratos, 1995a, op. cit.

[25] N. Alexandratos, 1995b, 'The outlook for world food and agriculture to year 2010', in Islam, op. cit., pp. 25–48

[26] Mitchell and Ingco, op. cit.

[27] Alexandratos, 1995a, op. cit.

[28] World Bank, 1994, *Global Economic Prospects and the Developing Countries*, Washington DC, World Bank

[29] Rosengrant *et al.*, op. cit.

[30] Peter Hazell, pers. comm.

[31] Alexandratos, 1995a, op. cit.

[32] Rosengrant *et al.*, op. cit.

3 A Doubly Green Revolution

We, Ministers, Heads of Agencies, and Delegates representing the membership of the Consultative Group on International Agricultural Research (CGIAR) . . .

Convinced that the new knowledge and technologies generated by scientific research are necessary to meet the rising food demand in a long-term sustainable way, from a limited and fragile resource base . . .

Call for the renewal and reinforcement of . . . work, aimed now at the multiple challenges of increasing and protecting agricultural productivity, safeguarding natural resources, and helping to achieve people-centred policies for environmentally sustainable development.

— The Lucerne Declaration and Action Plan [1]

It would be easy, on one reading of the model forecasts I described in the previous chapter, to be sanguine about the prospects for feeding the world in the next century. Market forces, the models imply, will create appropriate demand which, in turn, will stimulate the necessary technological innovation and infrastructural investment to ensure a sufficient supply of food. Many developing countries will need to purchase food, sometimes in large quantities, from the developed countries, but economic growth will provide the necessary purchasing power.

I do not believe this is satisfactory, for a number of reasons.

First, such a view of the world ignores the large proportion — one-fifth — of the developing-country population who are already hungry and whose plight is unlikely to be significantly improved, and it adds to them a new generation who will be hungry by the year 2020. I do not believe that hunger and poverty are inevitable conditions. The experience of the Green Revolution, which I describe in the next chapter, has amply demonstrated our ability to transform people's lives for the better. Hunger and poverty can be eliminated through the application of modern science and technology, provided they are used widely and are supported by appropriate economic and social policies, and a will to act.

Even if a limited level of hunger and poverty was deemed politically tolerable, the scale of deprivation predicted for South Asia and Sub-Saharan Africa is surely unacceptable. It will take a long time before African countries can generate sufficient foreign exchange to purchase the large amounts of food necessary to close their food gap. The real prices of Africa's traditional export crops are low, and the non-agricultural sector is small. It is also unlikely that African governments will be able to count on enough food aid to make up the difference. Poverty and hunger will increase rapidly in the coming years, unless something new is done. South Asia will fare better, but the food gap is substantial. Although exports of manufactured goods and GDP are likely to rise more rapidly in South Asia than in Sub-Saharan Africa, there will still not be sufficient foreign-exchange earnings to purchase the volume of cereals needed to ensure everyone has an adequate diet, even if it were available on the market.

Second, the model predictions depend on a significant rate of economic growth in the developing countries, a rate, as the World Bank admits, greater than in recent years and highly dependent on the effectiveness of current economic reforms. I cannot comment on the reliability of these forecasts. Experience from the first five years of the 90s indicate that, at least in Asia and Sub-Saharan Africa, growth rates are fulfilling the predictions, with rates in 1995 of about 8 per cent for Thailand and Indonesia and 9–10 per cent for Vietnam, South Korea and China, 5 per cent for Pakistan and Bangladesh and 6 per cent for India [2]. The overall growth rate for Africa was 3 per cent, with several countries exceeding 4 per cent, but there is continuing concern over the political stability of many countries, including some of those, such as Nigeria, containing the largest populations. Also of relevance to the optimistic scenario is the future economic prosperity of Eastern Europe and the former Soviet Union. The models assume the current economic reforms will be effective, predicting modest increases in GDP per capita (1–2 per cent) and agricultural production moving into surplus. However, the lack of infrastructure and the institutional obstacles to reform are formidable. If these are not overcome this region could remain a significant importer of grain, competing with the developing countries for the exports of North America and Western Europe. On the other hand, the potential for increased grain production in countries such as Russia, the Ukraine and Kazakhstan appears considerable and it is possible to envisage a scenario where they eventually become major suppliers of the developing countries' needs.

My third reason for dissatisfaction with the model predictions is their assumption of continuing increases in crop yields and production in line with recent trends. There are a number of grounds for questioning this assumption. Recent data on crop yields and production, which I discuss in Chapter 7, suggest a degree of stagnation which is worrying. There is widespread evidence of declines in the rates of yield growth. There are also data indicating greater variability in production in some regions and evidence, albeit largely anecdotal, of increasing production problems in those places where yield growth has been most marked.

The causes of this slowing in yield growth are not clear, although one factor is likely to be the cumulative effect of environmental degradation, partly caused by agriculture itself. The litany of loss is familiar. Soils are eroding and losing their fertility, precious water supplies are being squandered, rangeland overgrazed, forests destroyed and fisheries overexploited (Chapters 11 to 14). And, as I describe in Chapter 6, agricultural practices have become a significant contributor to the global pollutants that affect the ozone layers and produce global warming. These changes are already beginning to have significant adverse consequences on agricultural production. There is also clear evidence of instances where pesticides, far from solving pest problems, make them worse (Chapter 11). And, as we have long known, pesticides and nitrate fertilizers pose serious health hazards. Agriculture can expect increasing restrictions on its activities, along the lines of those already imposed on industry, that will limit the use of some of the inputs and practices which have delivered the recent high levels of production.

These concerns, I believe, add up to a formidable challenge. If, over the next two to three decades, we are to provide enough food for everyone we will have to:

– Increase food production at a greater rate than in recent years;
– Do this in a sustainable manner, without significantly damaging the environment;

and

– Ensure it is accessible to all.

It is a daunting prospect, whose magnitude becomes clear when we examine two contrasting scenarios of how this goal may be achieved:

Under *the first scenario*, the developed countries continue to produce food well in excess of their requirements and export this excess to meet the demand of the developing countries. If, as the models assume, environmental constraints to increased food production can be overcome and if the food needs of the poor are ignored, then there is little cause for concern. The demands of the developing countries, as expressed in national and international markets, will be met by national production in the areas of proven potential – the Green Revolution lands – and by imports via trade or aid from the developed countries. On the IFPRI model forecast, this would entail some 190 million tons of cereals being sold to the developing countries by the developed world in 2020.

However, if the food needs of the poor are not ignored, then under this scenario, a further 350 million tons would be required in 2020 as subsidized or free food aid (assuming a global requirement for the undernourished of 3,000 calories per person per day). This is equivalent to over thirty times the current supply of direct food aid and would be extremely costly. Such massive food aid would place heavy burdens on both the donors and the recipients. To meet their own market demand and that of the developing countries, and provide the necessary food aid, the developed countries would have to nearly double food production by the year 2020, from today's 860 million tons. This would require considerable increases in yields per hectare and the bringing back into production of all the land under cereals in the mid-1980s.

Inevitably, the environmental costs of such a scenario would be high. Nitrogen fertilizer use in the developed countries would increase, resulting in unacceptable levels of nitrate in drinking water. The costs to the developing countries also would be significant. Receipt and distribution of food aid requires considerable investment in infrastructure and administration. More important, the availability of free or subsidized aid in such large quantities will depress local prices and add to existing disincentives for local food production (Chapter 15).

These issues raise doubts about the viability of such a scenario, but a more fundamental objection is to an assumption implicit in the scenario – that a large proportion of the population in the developing world would fail to participate in global economic growth. *An alternative scenario*, which explicitly addresses this objection, is for the developing countries to undertake an accelerated, broad-based growth, not only in food production, but in agricultural and natural resource development, as part of a larger development process aimed at meeting most of their own food-production needs, including the needs of the poor. (Agriculture and natural resource use are inextricably related. In the rest of this book, agricultural development is understood to mean development of agriculture and natural resources, including forestry and fisheries.)

One justification for this scenario is the close link between economic and agricultural growth. The models I described in the last chapter treat economic growth as an input, driving agricultural development, whereas experience has amply demonstrated the power of agriculture as an engine for economic development. Increased production and employment in agriculture and natural resources can generate, directly or indirectly, considerable employment, income, and growth in the rest of the economy [3]. Very few countries have experienced rapid economic growth without preceding or accompanying growth in agriculture [4]. This is not because agriculture has a capacity for very fast growth, but because of its size. In the least-developed countries the agricultural sector typically accounts for over 80 per cent of the labour force and 50 per cent of the GDP and even modest rates of growth have a considerable multiplier effect, increasing rural incomes, which in turn create consumer demand and hence growth in the non-agricultural sector. The slope of the line in Figure 3.1 (which ignores the outliers, Singapore, Korea, Philippines, Burma) suggests that for each 1 per cent acceleration in agricultural growth there is about a 1·5 per cent acceleration in non-agricultural growth.

The relationship is not straightforward, however, as the graph indicates. The Philippines has failed to capitalize on the rapid agricultural growth generated by the Green Revolution [5]. Even when agricultural innovation and development stimulate economic growth, this does not necessarily lead directly to a reduction in poverty. Much depends on the nature of the innovations and how broad-based is the agricultural development they generate. While the introduction of irrigation and new varieties can create employment and incomes, certain kinds of mechanization associated with agricultural intensification can destroy jobs (see Chapter 6). For equitable economic growth, agricultural innovation needs to be deliberately focused on increasing production while, at the same time, creating employment both in agriculture and in related, rural-based industry.

Implicitly, this scenario recognizes that food security is not a matter solely of producing sufficient food. It is too simplistic to add up a nation's food production and divide by the size of population. Nor is it enough to

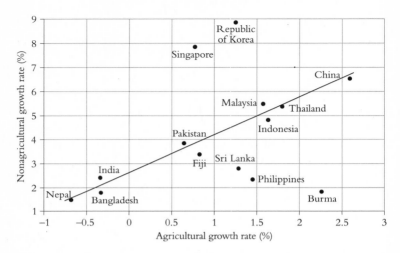

Figure 3.1 Per capita growth rates of agricultural and non-agricultural GDP in Asia [6]

point to declining food prices. A nation is food-secure only if each and every one of its inhabitants is food-secure, that is, has access at all times to the food required to lead a healthy and productive life (Chapter 15). To achieve this, each individual or, in practice, each household must grow sufficient food or be able to purchase the food from income earned either through selling agricultural products or by engaging in agricultural or non-agricultural employment. For urban dwellers, the only option is to engage in non-agricultural employment, but for the vast numbers of rural poor, if they are not growing enough food to meet their needs, they must have the means to purchase the food they require. For them, food security depends as much on employment and incomes as it does on food production, and agricultural and natural-resource development are crucial in both respects.

With surprisingly few exceptions, developing-country farmers, particularly in resource-poor environments, do not rely exclusively on farming. Their aim is to secure a livelihood for themselves and their families and to achieve this they usually pursue a range of productive activities, not all of which involve agriculture directly. Indeed, the ways in which a rural livelihood may be obtained are almost innumerable (Chapter 9). A household may depend on crop or livestock production, on exploiting natural resources, or on employment on or off the farm or, most commonly, on some combination of these. In practice, rural families decide on livelihood goals and then determine the optimal mix of activities, depending on their economic and social environment and the skills and assets at their disposal (Chapter 15).

Food security is usually but one goal; others are income and employment security and, linked to them, family size. Fertility rates depend on a complex of factors and, as I described in the previous chapter, there is evidence of the key role played by food and income security. The more confident women are about the immediate and longer-term future, the more likely they are to produce fewer children. Enhanced earning opportunities for women, as provided by the production, processing and trading activities generated by broad-based agricultural and natural-resource development, can contribute to lower fertility rates. The greater the degree of security and the higher the level of their education, the more will women take advantage of new opportunities and plan ahead for themselves and their families.

In addition, appropriate agricultural and natural-resource development can significantly contribute to greater environmental protection and conservation. Properly designed, sustainable approaches to food production and to forestry and fishery management can reverse land degradation, reduce pollution from agrochemicals, remove pressure on national parks and reserves, conserve biodiversity and, at the same time, increase food security.

Finally, there is a belief that vigorous agricultural and economic growth can stimulate world trade, providing significant benefits for all countries, developed and developing. The share of world exports going to the developing countries has grown from 13 per cent in the early 1970s to more than 26 per cent in the 1990s. As the developing countries prosper, they import more – a $1 increase in their agricultural growth is estimated to lead to an increase in the value of their imports of 73 cents [7]. And this means greater prosperity for the developed countries – the exports of the developed countries to the developing are currently valued at nearly $730 billion ($197 billion in the USA, where for every $1 billion of exports 20,000 jobs are created). It is this evidence of eventual mutual prosperity – what Per Pinstrup-Anderson and his colleagues at IFPRI have referred to as a 'win–win proposition' – that has driven the agreements reached under the Uruguay Round of trade negotiations, which aim to reduce protection and subsidies and open the world's markets to free trade (see Chapter 15). In the short term they may create difficulties, especially for those developing countries where strong protection has become institutionalized, and it will be essential for arrangements to be made to smooth the transition and to ensure the condition of the poor is not worsened. But, in the long run, the expectation is that freer trade will provide widespread incentives and opportunities from which the poor will significantly benefit. Greater all-round prosperity and fuller inclusion will, it is hoped, reduce conflict and increase stability, creating a further basis for equitable progress.

In summary, a major investment in agriculture and natural resources in the developing countries could:

– Create employment and incomes for the mass of the poor;
– Deliver food security;
– Help to reduce birth rates through increased food and income security;
– Protect and conserve the environment;
– Stimulate development in the rest of the economy;
– Ensure prosperity in the industrial world through the stimulation of global trade; and
– Increase the likelihood of political stability.

The challenges posed by this scenario are considerable. There is no single recipe for successful agricultural development, though there is a broad consensus on many of the essential ingredients. These include:

– Economic policies that do not discriminate against agriculture, forestry or fisheries [8];
– Liberalized markets for farm inputs and outputs with major private-sector involvement;
– Efficient rural financial institutions, including adequate access by all types of farmers to credit, inputs and marketing services;
– In some cases, land reform or redistribution;
– Adequate rural infrastructure, including irrigation, transport and marketing;

– Investments in rural education, clean water, health, nutrition programmes and family planning;
– Specific attention to satisfying the needs of women and of ethnic and other minority groups and securing their legal rights; and
– Effective development and dissemination of appropriate agricultural technologies in partnership with farmers.

Although listed separately in this manner, they are intimately interconnected. While economic liberalization within developing countries and reform of international trading policies are necessary prerequisites for significant agricultural growth, they are not sufficient. Accelerated growth in agricultural output cannot be maintained without adequate investments in rural infrastructure and in agricultural research and extension. Indeed, without such investment the results of liberalization policies may well fall short of expectations and set governments against market-oriented approaches.

And they will not contribute significantly to poverty alleviation and the reduction in inequity, at least in the short term, unless the poor are deliberately targeted. Essential steps include the creation of employment for the land–poor and landless, increased production on small and medium-sized as well as large farms, provision of nearby input and output markets [9], and, recognizing where the rural poor are mostly located, attention to regions of lower agroclimatic and resource potential, not just the best [10].

This means that agricultural innovation, at least in the developing countries, cannot be left simply to market forces. In the industrialized countries the production of new varieties and agricultural chemicals has been assigned increasingly to the private sector. Better-off farmers, often heavily subsidized, are able to afford the products of expensive research. Private companies can patent and protect their discoveries for sufficient time to realize an acceptable profit. But inevitably, private research focuses on the major high–value crops, on labour-saving technologies and on the needs of capital-intensive farming. By contrast, research to feed the poor is less attractive. It frequently involves long lead times, for example in developing new plant types of minor staples. It is risky, particularly when focused on heterogeneous environments that are subject to high climatic or other variability. And the beneficiaries have little capacity to pay for the research. The products cannot be restricted to those who pay and intellectual property rights can rarely be protected.

As the history of the Green Revolution makes clear (see the next chapter), publicly funded research, if deliberately aimed at low-cost food production, has benefits that spread to farmers, large and small, to other rural dwellers and, most important, to poor consumers. If well designed and targeted, it can exploit the potential for positive externalities, especially as they benefit the poor. It also has a growing role to play in ensuring that technologies are environmentally appropriate. Inevitably, the beneficiaries of such technologies are often not the users, or at least not the users alone, and are rarely well served by private investment.

Investments in agricultural research continue to give consistently high rates of return. This has been demonstrated time and time again in cost/benefit analyses of individual research projects and programmes (Table 3.1). Runs of the IFPRI model based on a reduction in international aid for agricultural research in the developing countries of $1·5 billion a year result in a 10 per cent fall in their cereal production [11]. Prices rise by between 25 and 40 per cent, the availability of food is only marginally better than today and the number of malnourished children increases to 205 million. On the other hand a rise in such aid by $750 million will increase production by 6 per cent, prices will not grow and the number of malnourished children will fall to 106 million.

Table 3.1 Returns from publicly funded agricultural research and extension [12]

Country	Research target	Years	Rate of return (%)
Bangladesh	Wheat and rice	1961–77	30–35
Brazil	Soybeans	1955–83	46–69
Brazil	Irrigated rice	1959–78	83–119
Chile	Wheat and maize	1940–77	21–34
Colombia	Rice	1957–64	75–96
Mexico	Wheat	1943–63	90
Pakistan	Wheat	1967–81	58
Peru	Maize	1954–67	50–55
Philippines	Rice	1966–75	75
Rwanda	Potato seed	1978–85	40
Senegal	Cowpeas	1981–7	63

There is also a considerable direct benefit from developing-country agricultural research for the developed countries. Research undertaken as part of the Green Revolution on dwarfing of cereals and on resistance to diseases, such as rust, has helped increase and stabilize cereal production in North America and Europe. According to the detailed analysis of the US grainlands carried out by IFPRI staff, by the early 1990s about one-fifth of the total US wheatland was being sown to varieties derived wholly or in part from varieties developed at the International Centre for Corn and Wheat Improvement (CIMMYT) in Mexico, and for the California spring-wheat crop this was nearly 100 per cent [13]. And 73 per cent of the riceland in the USA was being sown to varieties derived in part from varieties developed by the International Rice Research Institute (IRRI) in the Philippines. Since the early 1960s the USA has invested $71 million in CIMMYT and $63 million in IRRI; the calculated returns have been up to $13 billion from the former and up to a billion from the latter.

I believe the arguments I have presented in this chapter, when taken together, point to the need for increased public investments in agricultural research – in effect for a second Green Revolution, yet a revolution that does not simply reflect the successes of the first. The technologies of the first Green Revolution were developed on experiment stations that were favoured with fertile soils, well-controlled water sources, and other factors suitable for high production (see the next chapter). There was little perception of the complexity and diversity of farmers' physical environments, let alone the diversity of the economic and social environment. As Randolph Barker and his colleagues put it, 'Unfortunately, scientists frequently saw their responsibility as ending at the experiment station gate' [14]. The new Green Revolution must not only benefit the poor more directly, but must be applicable under highly diverse conditions and be environmentally sustainable. By implication, it must make greater use of indigenous resources, complemented by a far more judicious use of external inputs.

In effect, we require a Doubly Green Revolution, a revolution that is even more productive than the first Green Revolution and even more 'green' in terms of conserving natural resources and the environment [15].

Over the next three decades it must aim to:

- Repeat the successes of the Green Revolution;
- On a global scale;
- In many diverse localities;

and be:

41

– Equitable;
– Sustainable; and
– Environmentally friendly.

While the first Green Revolution took as its starting point the biological challenge inherent in producing new high-yielding food crops and then looked to determine how the benefits could reach the poor, this new revolution has to reverse the chain of logic, starting with the socio-economic demands of poor households and then seeking to identify the appropriate research priorities. Its goal is the creation of food security and sustainable livelihoods for the poor.

Success will not be achieved either by applying modern science and technology, on the one hand, or by implementing economic and social reform on the other, but through a combination of these that is innovative and imaginative. It will require a concerted effort on the part of the world community, in both the industrialized and developing countries, the application of new scientific and technological discoveries in a manner that is environmentally sensitive and above all, as I discuss in Chapter 10, the creation of new partnerships between scientists and farmers that will respond to the needs of the poor. In the following chapters I expand on the justifications for this new concept and elaborate on what is needed to ensure success.

Notes

[1] CGIAR, 1995, *Renewal of the CGIAR: Sustainable Agriculture for Food Security in Developing Countries*, Washington DC, Consultative Group on International Agricultural Research

[2] IMF, 1995, *World Economic Outlook, October 1995*, Washington DC, International Monetary Fund

[3] P. Hazell and S. Haggblade, 1993, 'Farm–nonfarm linkages and the welfare of the poor', in M. Lipton and J. van der Gaag (eds.), *Including the Poor*, Washington DC, World Bank

[4] J. W. Mellor, 1995, 'Introduction', in J. W. Mellor (ed.), *Agriculture on the Road to Industrialization*, Baltimore, Md, Johns Hopkins University Press, pp. 1–22

[5] R. M. Bautista, 1995, 'Rapid agricultural growth is not enough: the Philippines, 1965–80', in Mellor, op. cit., pp. 113–49

[6] Mellor, op. cit.; using data from World Bank, 1989, *World Tables*, Baltimore, Md, Johns Hopkins University Press

[7] P. Pinstrup-Anderson, M. Lundberg and J. L. Garrett, 1995, *Foreign Assistance to Agriculture: A Win—Win Proposition*, Washington DC, International Food Policy Research Institute (Food Policy Report)

[8] World Bank, 1992, *The Political Economy of Agricultural Pricing Policy*, Baltimore, Md, Johns Hopkins University Press

[9] H. P. Binswanger, K. Deininger and G. Feder, 1995, 'Power, distortions, revolt and reform in agricultural land relations', in J. Behrman and T. N. Srinivasan, *Handbook of Development Economics*, Vol. IIIB, Amsterdam (Netherlands), Elsevier, pp. 2661—772

[10] M. and M. Lipton, 1993, 'Creating rural livelihoods: some lessons for South Africa from experience elsewhere', *World Development*, 21, 1515—48; P. Hazell and J. L. Garrett, 1996, 'Reducing poverty and protecting the environment: the overlooked potential of less-favoured lands', *2020 Brief*, 39, Washington DC, International Food Policy Research Institute

[11] M. W. Rosengrant, M. Agcaoili-Sombilla and N. D. Perez, 1995, *Global Food Projections to 2020: Implications for Investment*, Washington DC, International Food Policy Research Institute (Food, Agriculture and the Environment Discussion Paper 5)

[12] R. G. Echeverria, 1989, 'Assessing the impact of agricultural research', in ISNAR, *Methods for Diagnosing Research System Constraints and Assessing the Impact of Agricultural Research*. Vol. II, *Assessing the Impact of Agricultural Research*, The Hague (Netherlands), International Service for National Agricultural Research

[13] P. G. Pardey, J. M. Alston, J. E. Christian and S. Fan, 1996, *Hidden Harvest: US Benefits from International Research Aid*

[14] R. Barker, R. W. Herdt and B. Rose, 1985, *The Rice Economy of Asia*, Washington DC, Resources for the Future

[15] G. R. Conway, U. Lele, J. Peacock and M. Piñeiro, 1994, *Sustainable Agriculture for a Food Secure World*, Washington DC, Consultative Group on International Agricultural Research/Stockholm, Swedish Agency for Research Cooperation with Developing Countries

4 Past Successes

The Japanese have made the dwarfing of wheat an art. The wheat stalk seldom grows longer than 50 to 60 centimeters. The head is short but heavy. No matter how much manure is used, the plant will not grow taller; rather the length of the wheat head is increased. Even on the richest soils, the wheat plants never fall down.

                      – US adviser to the Meiji government, 1873 [1]

Without the Green Revolution, the numbers of poor and hungry today would be far greater. Thirty-five years ago, according to FAO, there were about 1 billion people in the developing countries who did not get enough to eat, equivalent to 50 per cent of the population, compared to the under 20 per cent today [2]. If the proportion had remained unchanged the hungry would be in excess of 2 billion – more than double the current number. The achievement of the Green Revolution was to deliver annual increases in food production which more than kept pace with population growth.

Many factors contributed to this success story, but of central importance was the application of modern science and technology to the task of getting crops to yield more. Cereal yields, total cereal production and total food production in the developing countries all more than doubled between 1960 and 1985. Over the same period their population grew by about 75 per cent. As a result, the average daily calorie supply in the developing countries increased by a quarter, from under 2,000 calories per person in the early 1960s to about 2,500 in the mid-80s, of which 1,500 was provided by cereals (Figure 4.1) [3].

The history of the Green Revolution is well known, but is worth recounting here as a reminder of the power and limitations of innovative technology, and the crucial importance to its success of the economic, social and institutional environment within which it has to operate.

A careful analysis of the trends in agricultural productivity in a variety

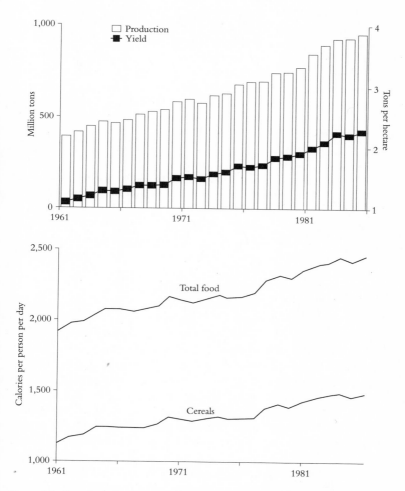

Figure 4.1 Cereal production, cereal yields and calorie supply in the developing countries, 1961−86

of countries, both developed and developing, suggests there is a point in history when yields begin to take off [4]. While agricultural production remains based on traditional practices, with fewer or no outside inputs, yield gains are modest, of the order of 1 per cent or less per annum. Production grows, barely perceptibly and largely as a result of increases in cropped area, until a critical threshold is reached when a transition to a new basis for production occurs. Much-improved varieties appear, farmers turn to inorganic fertilizers and synthetic pesticides, invest in irrigation and drainage and adopt a range of other new technologies, including the purchase of agricultural machinery.

From their analyses, Cornelius de Wit and his colleagues of the University of Wageningen in the Netherlands conclude this transition occurs when cereal yields reach about 1,700 kg/ha. Below the take-off point yield increases are of the order of 17 kg/ha per year; afterwards they are 50–85 kg/ha. There are, of course, exceptions: some countries (China is an example) experienced take-off from much lower yield levels, others, such as Britain and Japan, produced several spurts of growth with relatively low rates in between. Selection of indigenous rice varieties in Japan, coupled with increased irrigation, resulted in a yield increase between the 1880s and the First World War of about 40 kg/ha per year [5]. Yields grew more slowly between the wars and then took off again after the Second World War, with annual yield growth of 75 kg/ha per year.

In most developed countries the take-offs occurred soon after the end of the Second World War. High-yielding varieties, new fertilizer formulations and more effective pesticides came on the market, together with machinery that permitted more timely agricultural operations. Economic incentives, which included guaranteed prices, deficiency payments and other forms of subsidy, made sure the new technologies were widely adopted. Thereafter, growth in production was nearly entirely due to yield increase; in most developed countries the cropped area has declined since 1950.

The take-offs in the developing countries mostly occurred at the end of the 1960s as the new varieties produced by the Green Revolution began to be widely adopted. William Gaud, the administrator of USAID, first coined the name 'Green Revolution' [6]. At the time it was an appropriate description of a momentous event. Today 'Green' signifies the environment; then the image it conveyed was of a world covered with luxuriant and productive crops – the green swathes of young wheat- and ricefields.

It was truly also a revolution in the scale of the transformation it achieved although, as we shall see, it did not go nearly far enough.

The origins of the revolution lay in a joint venture, the Office of Special Studies, established by the Mexican Ministry of Agriculture and the Rockefeller Foundation in 1943 [7]. At the time, Mexican grain yields were very low, maize averaging about a quarter of US yields and wheat yielding less than 800 kg/ha, even though most of the wheatland was irrigated. The Office was headed by George Harrar, with Edwin Wellhausen, a maize breeder, Norman Borlaug, a plant pathologist, and William Colwell, a soil scientist. Eventually the office was to have twenty-one US and 100 Mexican scientists, mostly working at an experiment station at Chapingo on the rain-fed central plateau. Its remit was to improve the yields of the basic food crops, maize, wheat and beans.

The research programme concentrated first on maize, the mainstay of the Mexican diet, consumed in the form of the thin, flat, unleavened bread called the tortilla. Mexican agronomists had already discovered that most strains of maize grown in the United States were not well adapted to Mexican conditions, so the programme set out to try to duplicate the US achievement of breeding high-yielding, hybrid maizes but using indigenous varieties as a basis. Maize is a cross-pollinating crop, so that the seed collected by the farmer from his or her crop at the end of the season is usually highly variable. More uniform and higher-yielding hybrids can be created by deliberately crossing two distinct lines which have been inbred through several generations by self-pollination. The resulting hybrids combine the best features of both lines and usually have added a certain hybrid vigour [8]. At the time the Mexican programme began, about half of the cornland in the USA was planted to hybrids. However, the disadvantage of hybrids is that new seed for each season has to be produced by repeating the cross, since the seed gathered at the end of the season from the hybrid crop will have lost the hybrid vigour and become contaminated. An alternative approach, adopted by Edwin Wellhausen, was to grow four or more inbred lines in an isolated field and let them cross naturally. These so-called 'synthetics' yielded better than the best existing varieties by 10–25 per cent and the farmers could simply save seed from their best plants from year to year [9].

Success came very quickly. In 1948 1,400 tons of seed of the improved maize varieties were planted. The new seed, good weather that season and the ready availability of fertilizer resulted in a record harvest and, for the first time since the revolution of 1910, Mexico had no need of imports. By the 1960s over one-third of Mexico's maizeland was being planted to new high-yielding varieties and maize yields were averaging over 1,000 kg per hectare. Total production had increased from 2 to 6 million tons.

Although maize was the main staple crop, Mexico was importing about a quarter of a million tons of wheat per year. Yields were very poor – 'most varieties were a hodge-podge of many different types, tall and short, bearded and beardless, early ripening and late ripening. Fields usually ripened so unevenly that it was impossible to harvest them at one time without losing too much over-ripe grain or including too much under-ripe grain in the harvest' [10]. In northern and central Mexico the soils had lost most of their fertility. On the newer, well-irrigated lands in the Pacific north-west the soil was generally fertile enough to produce high yields but stem rust was very destructive. Epidemics in three consecutive years, 1939–41, in Sonora had caused many farmers to reduce their wheatland or stop growing the crop altogether.

The wheat programme, under the direction of Norman Borlaug, began by testing over 700 native and imported wheat varieties for rust resistance [11]. While some of the imported varieties were both more resistant to rust and higher yielding than the Mexican varieties, they had the disadvantage of late ripening. They needed the longer days of the northern summers to mature. But by crossing the imports with the best Mexican varieties, Borlaug produced, in 1949, four rust-resistant varieties each adapted to a particular ecological region of Mexico. Wheat is a self-pollinating crop, so that crosses have to be made by hand, but once made the new varieties will breed true and farmers can use their harvested seed for the next year's crop. By 1951 the new varieties were being grown on 70 per cent of the total wheatland and, five years later, Mexico was producing over a million tons of wheat, with an average national yield of 1,300 kg/ha. Imports of foreign wheat were no longer required.

The next step in the wheat programme was to improve yields through greater use of fertilizers. Experiments on properly irrigated soils showed that 140 kg/ha of nitrogen raised yields more than fourfold. Even on rain-fed soils, yields more than doubled and the addition of phosphate produced five- or sixfold increases. But the traditional varieties had a tendency to grow tall and lodge, that is they fell down under the weight of the luxuriant green growth produced by the extra nutrient uptake, in effect making the grain liable to rot and be less harvestable. The breeders also recognized the need for non-shattering varieties that would hold the grains until they were ripe enough for mechanized harvesting and threshing. And wheats with better milling and baking quality were required, now that Mexico was relying solely on its own wheats and not blending with the stronger, imported varieties.

As the quotation at the beginning of this chapter testifies, short-strawed varieties had long existed in Japan. In 1935 the Japanese had produced a new dwarf, Norin 10, by crossing one of their traditional dwarfs with Mediterranean and Russian varieties imported from the USA [12]. This was spotted by a US agricultural officer working in Japan in 1946 and seeds were sent to Washington State University. Initially the crosses with US wheats resulted in sterile offspring but then a fertile cross was made from which, ten years later, Orville Vogel, of the US Department of Agriculture (USDA), produced a new semi-dwarf variety called Gaines that yielded a world record of over 14 tons/ha.

Norman Borlaug heard of Vogel's work and in 1953 obtained some of his early breeding material to cross with traditional Mexican varieties. By growing two crops a year, a summer crop at Chapingo and a winter crop at a second experiment station in the state of Sonora, lying in the rich, irrigated plain of the Pacific north-west of Mexico, he produced, in record time, two new, superior dwarf-wheat varieties. These were adapted to a wide range of day-length and other environmental factors, and highly responsive to fertilizer applications (Figure 4.2). By 1966 the new varieties were yielding 7 tons/ha. A decade later, virtually all of Mexico's wheatland was under these varieties and the country's average yield was close to 3 tons/ha, having quadrupled since 1950 (Figure 4.3). By 1985 total production had increased to 5·5 million tons.

Following the success of the new wheat and maize varieties in Latin America, attention turned to the needs of Asia. Over much of Asia the staple diet was, and still is, rice. Unlike wheat, rice is mostly grown by smallholders (on farms of less than 3 ha) for home consumption. In the 1950s, over 90 per cent of the rice produced in the world was grown in Asia; national yields averaged between 800 and 1,900 kg/ha [13].

Early in this century, Japan, China and Taiwan had developed a number of new rice varieties which had led to significant increases in production, but the major impact on Asian rice yields came in the 1960s as a result of the skilful efforts of two groups of breeders, working in ignorance of each other, in China and the Philippines. The benefits of scientists being able

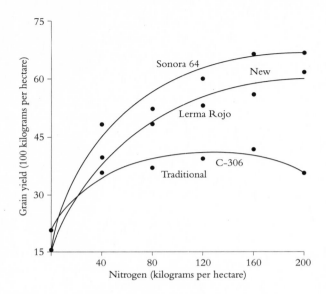

Figure 4.2 Responses of new and traditional wheats to fertilizers [14]

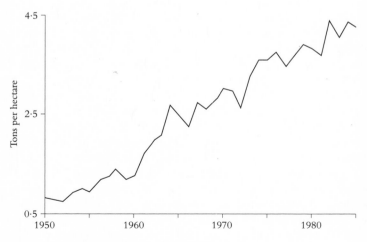

Figure 4.3 Growth in wheat yields in Mexico

to pursue clear goals in multidisciplinary teams had convinced those involved in the Mexican programme of the need to create purpose-built research institutes which would attract scientists of the highest calibre. In 1961 the International Rice Research Institute (IRRI) was established at Los Banos in the Philippine province of Luzon as a joint venture between the Philippines government and the Ford and Rockefeller Foundations. This was the first of a family of new research institutes equipped with first-class laboratories and adjoining experimental plots located on good, irrigated land [15]. Excellent living conditions and international salaries added to the ability to attract the best scientists from around the world.

The Ford Foundation became a partner partly because its community development programme in India, started in 1951, had underlined the importance of agricultural research. The programme, largely run by social scientists, had assumed that improved technology was readily available and needed only a programme of vigorous education for it to be implemented. However, the village extension workers often proved inexperienced in agriculture and, more important, encouragement of increased fertilizer use turned out to be ineffective because the traditional cereal varieties lodged. Forrest Hill, an agricultural economist and vice-president of the foundation, concluded that the foundation had 'got the cart before the horse'. What was needed, he believed, was innovative agricultural research to support the extension work.

The first director of IRRI was Robert Chandler. He assembled a team of rice experts drawn from the USA, India, Japan, Taiwan, Ceylon and the Philippines [16]. Experience in Mexico with the wheat programme and the knowledge already obtained from breeding programmes in India provided a blueprint for the new rices that were required (Box 4.1). A large collection of rice types was quickly amassed at Los Banos and of the crosses made in 1962, a particularly promising combination was between the tall, vigorous variety Peta from Indonesia and Dee-geo-woo-gen, a short, stiff-strawed variety from Taiwan which contained a single recessive gene for dwarfing. In 1966 the new variety, named IR8, was released for commercial planting in the Philippines. It was immediately successful and, amid considerable publicity, was dubbed the 'miracle rice'.

In parallel with the work at IRRI a very similar breeding programme had been launched at the Academy of Agricultural Sciences in the Chinese province of Guandong [17]. Although this was only known much later, the Chinese and Philippine teams were using breeding material containing

Box 4.1 **A blueprint for the new rice varieties** [18]

— A short, stiff stem (90–110 cm), giving resistance to lodging

— Erect, narrow leaves, resulting in increased efficiency of sunlight utilization

— High tillering and a grain to straw ratio of 1:1, producing high fertilizer responsiveness

— Time of flowering insensitive to day-length, giving flexibility in planting date and location

— Early maturity (less than 130 days), giving increased output per hectare per day

— Resistance to the most serious pests and diseases: stem borer and rice blast

— Wide adaptability in Asia

— Highly nutritious, with a high protein content and a better balance of amino acids

— High palatability

the same dwarfing gene – the Taiwanese variety Dee-geo-woo-gen had probably originated in southern China. In 1959 the Chinese produced their first successful cross, similar in many respects to IR8, and known as Guang-chai-ai. It was rapidly taken up in the province of Guandong and in Jiangsu, Hunan and Fujian. By 1965, a year prior to the release of IR8, it was being grown on 3·3 million hectares.

IR8 combined the seedling vigour of Peta with the short straw of Dee-geo-woo-gen. It was, like the new wheats bred in Mexico, highly responsive to fertilizer and essentially insensitive to photoperiod, maturing in 130 days. Under irrigation it yielded 9 tons/ha in the dry season and on the IRRI farm, when continuously cultivated with a rapid turn-around between crops, produced average annual yields of over 20 tons/ha. In Asian regional trials it outyielded virtually all other varieties, producing from 5–10 tons/ha.

There was an immediate impact on Philippine rice production. By 1970, one and a half million hectares, or half of the Philippines' riceland was planted to the new varieties and the yield take-off had occurred. The

Philippines became self-sufficient in rice production in 1968 and 1969 for the first time in decades, although this was temporarily lost in the early 1970s owing to bad weather and disease outbreaks. A decade later, 75 per cent of the riceland was planted to the new varieties, average yields were over 2,000 kg/ha and rising at nearly 70 kg/ha per year (Figure 4.4).

To begin with, the new varieties were distributed somewhat haphazardly. When IR 8 was first released, farmers who turned up at IRRI could have 2 kg of seed free, provided they left their name and address. The seed spread within a few months to over two-thirds of the provinces of the Philippines. Later, seed was distributed through government agencies and a newly formed Seed Growers Association, composed of private farmers who both grew and marketed the new seed. Standard Oil of New Jersey (ESSO) also established 400 agro-service centres in the Philippines to serve as marketing outlets not only for fertilizer but for seed, pesticides and farm implements [19].

Two Mexican farm advisers working in El Salvador had hit on the idea of putting together in one package all the basic inputs a farmer would need to try out a new variety on a small patch of ground. The idea quickly spread to other countries and was tried on a massive scale in the Philippines,

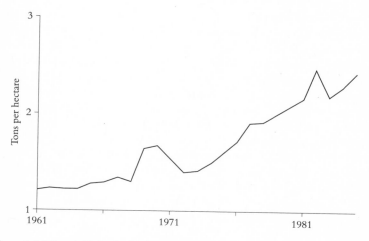

Figure 4.4 Rice yields in the Philippines, 1961–85

where a typical package contained 0·9 kg of IR8 seed, 19 kg of fertilizer and 2·7 kg of insecticide. The packages were produced by governments and also sold by fertilizer companies [20]. In addition to continuing its programme of food aid, the US aid agency, USAID, began to support fertilizer shipments in the 1960s and to finance rural infrastructure – farm-to-market roads, irrigation projects and rural electrification. It also funded a large force of technical assistance experts. Contracts were signed with US land-grant colleges to assist institution-building in education, research and extension and to create agricultural universities in the land-grant tradition. One of the most successful partnerships was between Cornell University and the University of the Philippines at Los Banos, next to the IRRI campus.

A conscious objective of the Green Revolution, from the beginning, was to produce varieties that could be grown in a wide range of conditions throughout the developing world. To meet this goal, the breeders in Mexico had successfully bred the new wheats to be photoperiod-insensitive, that is, they would flower and produce grain at any time of year, in contrast to traditional varieties which tend to flower at certain seasons, for example when the days are shortening. Provided the temperature was above a certain minimum and there was sufficient water, the new varieties would grow almost anywhere.

As early as the 1950s, successful trials of the new Mexican maize and wheat varieties were conducted in Latin America and Asia. The new varieties in India yielded at least a ton more than the local varieties. However, local pests and diseases and location-specific soil–management problems remained major constraints. And the importance of getting the package right was illustrated by an attempt to introduce the Mexican soft wheats to Tunisia, where the farmers are used to growing hard wheats, eaten in the form of couscous [21]. In the first year the trials gave average yields of 1·5 tons/ha, three times that of the hard wheat. The early maturity of the soft wheats also meant they could be harvested quickly before the drought set in. But the greatly expanded planting in the third year was a catastrophe, with yields of only 300 kg/ha. The seed was of poor quality and, because of a complicated bureaucracy, was distributed too late. The farmers also planted it too deep for fear of drought and failed to keep the weeds down. In the fourth year they reverted to the traditional hard–wheat varieties.

To oversee the international effort, a wheat and maize improvement programme was established in 1966 under the umbrella of the International Centre for the Improvement of Maize and Wheat, located at Chapingo in Mexico and known by the initials (from its Spanish name) CIMMYT. The new seeds were mostly exported at a price very little above the world market price. In 1967–8 Pakistan imported sufficient of the new wheat seeds to plant more than 400,000 hectares. Pakistan and India's yield take-offs occurred in 1967; subsequent yield increases were about 50 kg/ha per year (Figure 4.5). A major impetus for the rapid uptake of the new varieties in South Asia was the two consecutive failures of the monsoon in 1966 and 1967. The USA, which was the only country carrying food reserves of any size, had responded by shipping one-fifth of its grain crop to India [22]. This degree of dependence was recognized by both sides as being risky and undesirable. In some instances, food-aid agreements had permitted countries to put off serious agricultural development plans, and the large volume of food received had depressed prices and reduced the incentive for farmers to produce more. Then the monsoon failures helped to wipe out the world's surpluses and prices rose dramatically (rice going from $120 to over $200/ton in 1967). Faced with a food crisis of this magnitude, a number of developing-country leaders quickly recognized the potential of the new varieties. Presidents Ayub of Pakistan and Marcos of the Philippines and Prime Minister Demirel of Turkey took a personal and active part in the promotion of the Green Revolution, and were able to take some of the credit for its success. The re-election of President Marcos in 1969 owed much to the attainment of self-sufficiency in rice production.

The new wheats also proved popular in Argentina, where their early maturity made them appropriate for double-cropping with soybeans [23]. And in Egypt wheat yields took off in 1969, initially growing at over 60 kg/ha per year but after 1980 at 200 kg/ha [24]. The hybrid maizes were also successfully introduced into Egypt and into Kenya and Zimbabwe. However, the biggest impact came from their adoption in Latin America. To begin with this was slow because the traditional peasant farmers were reluctant to switch from the open-pollinated varieties which could be retained as seed on the farm for next year's harvest. However, by the early 1980s half the maize area of Latin America was sown to hybrids and yields had grown by a third since 1960. Chile, in particular, experienced dramatic growth, at over 200 kg/ha per year after take-off in 1964. According to Donald Plucknett of the CGIAR this is the highest sustained growth rate

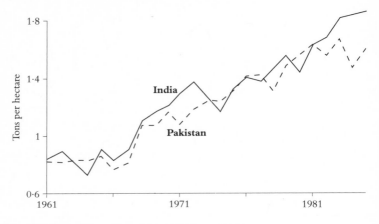

Figure 4.5 Growth in wheat yields in India and Pakistan

that any cereal has so far experienced [25]. Average yields in 1985 were approaching 6 tons/ha.

The uptake of the new rice varieties was similarly dramatic. Twenty tons of IR8 seed were sent to India in 1966 and a further 5 tons to other Asian and Latin American countries. In Colombia it gave rise to a number of local variants developed by the Centre for Tropical Agriculture (Centro Internacional de Agricultura Tropical – CIAT) which soon replaced the indigenous varieties. By 1985 average yields were approaching 5 tons/ha and rice had become the dominant Colombian food crop [26]. Another remarkable transformation occurred in Indonesia, where oil revenues were used to finance the adoption of the new varieties and their accompanying packages. The take-off occurred in 1967, with subsequent rice-yield increases of over 100 kg/ha (Figure 4.6).

In China rice yields gained an extra boost from the development of hybrid rice. Rice, like wheat, is self-pollinating but the Chinese developed an inexpensive technique permitting cross-pollination on a large scale. The resulting hybrid rices have the same qualities of hybrid vigour as the hybrid maizes: yields are some 20 per cent, or about a ton, greater than the semi-dwarf rices. The first hybrid rice was distributed in 1974 and within a few years it was being planted on 15 per cent of Chinese riceland. China also introduced new high-yielding wheats from Mexico and crossed them with local varieties. Grain yields grew steadily in the 1970s and then

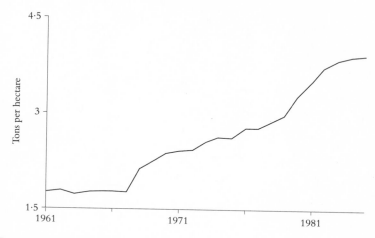

Figure 4.6 The growth of rice yields in Indonesia

accelerated after 1978, following the break-up of the communes, the re-establishment of the household as the unit of production and the encouragement of local markets (Figure 4.7) [27].

By the 1980s the Green Revolution and new Chinese varieties were dominating the grain lands of the developing world (Figure 4.8) [28].

Inevitably, there were many 'teething problems' in the early years. Governments were unprepared for the rapid rise in production. The land planted to IR8 in Pakistan increased a hundredfold, to over 400,000 hectares, in only a year. Storage, transport and marketing systems were sometimes overwhelmed. The 1968 Indian wheat harvest was one-third greater than the previous record and schools had to be closed in order to store the grain [29]. A huge harvest in Kalimantan, on the island of Borneo, went to waste because there was insufficient transport to get it to the centres of demand in Java. But these were the problems of success and they were quickly overcome.

Another early problem was the poor acceptability of some of the new varieties. Even poor, undernourished people retain a pride in eating good-quality grain. There was preference in India and Pakistan for chapatis made from the traditional white grains rather than the reddish grains of the new varieties, but this problem was soon solved by breeding for white

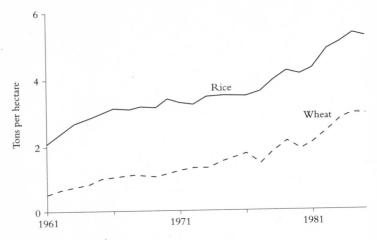

Figure 4.7 Growth of rice and wheat yields in China

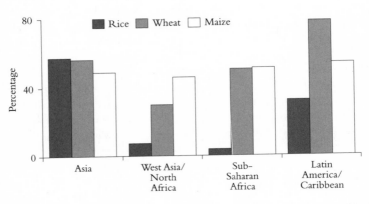

Figure 4.8 Proportion of cereal lands under new Green Revolution or Chinese varieties in the early 1980s [30]

grain colour [31]. The grain of the first IRRI rices, such as IR8, was also rejected because it tended to harden excessively after cooking and farmers received a lower price than for the traditional grains. Quality improved with later varieties, but many farmers continued to prefer the traditional taste. In Indonesia the growing of traditional rice varieties was prohibited, a ban sometimes enforced by the destruction of crops in the fields. Farmers responded by cultivating the traditional rices in their home gardens or on the margins of fields growing the new varieties.

Other problems were more persistent and less amenable to simple solutions. They often required changes in government policies and the creation of new agencies. Inevitably the solutions created yet further problems. An example was the credit needs of the new technology. Adoption of the high-yielding packages was expensive: in Bangladesh the cost of the necessary inputs was 60 per cent more than for the traditional varieties. Small, subsistence farmers, often tenants or sharecroppers, could only afford the new packages if they borrowed from local moneylenders, invariably at high rates of interest. In the Philippines, where the majority of farmers are tenants, cash was borrowed from the landlords at rates of 60–90 per cent per annum, often producing a permanent state of indebtedness. The government response was to set up an Agricultural Guarantee Loan Fund, established at the Central Bank. This, in turn, supported numerous private rural banks which loaned without collateral and at reasonable interest rates. It was an important factor in determining the very rapid uptake of the new varieties in the Philippines.

However, there were drawbacks. Under the traditional feudal system, the tenants provided personal services for the landlord, gathering fuel or lending a hand with house repairs. In return, landlords helped with rice or money at times of economic hardship, providing a degree of protection against the outside world [32]. With the advent of the Green Revolution, this relationship became more commercial. Both sides benefited: yields on tenant farms rapidly increased as farmers gained from the landlord's better access to technical information, machinery and inputs. But the inputs had to be paid for and in bad years there was no longer any latitude. Credit repayments under the government scheme were due on schedule and defaulters were punished. The banks were less accommodating than the landlords.

Governments usually intervened directly to assure the supply of inputs. Demand for fertilizers grew rapidly in the early years, sometimes outstripping supply and forcing up prices [33]. The response was to fix prices and provide generous subsidies. Under the BIMAS programme in Indonesia, quotas were awarded to licensed importers, prices were fixed and distributors and retailers appointed right down to the cooperatives at village level. By the mid–1980s the subsidy for fertilizers had reached 68 per cent of the world price, for pesticides 40 per cent and for water nearly 90 per cent. Such high levels of subsidy created serious environmental problems (see Chapter 6). Also because the distribution of subsidized inputs tended to remain in the hands of government or quasi-government agencies, corruption became widespread and, in some situations, institutionalized. This has been most evident in government irrigation schemes (see Chapter 13) [34].

Pest and disease outbreaks have been an especially severe consequence of the Green Revolution [35]. In some cases the cause-and-effect relationship is simple. Pest populations have grown in response to higher nitrogen applications and diseases have become more prevalent in the microclimate created by the densely leaved, short-strawed wheats and rices. But often it is a combination of factors – higher nutrient levels, narrow genetic stock, uniform continuous planting and the misuse of pesticides – that have created conditions which encourage pest and disease attack (see Chapter 11).

According to some critics of the Green Revolution, the growth in production owed little to the new varieties, and was primarily due to agricultural expansion, arable cultivation moving on to increasingly marginal lands. But the evidence is otherwise: although area increases were important in the 1950s, the subsequent gains were largely derived from increasing yield per hectare. Other commentators claim the growth in production has been due to infrastructural and institutional change rather than the specific technical innovations, pointing to the lack of signs of the impact of the new varieties in the regional data [36]. But this is because the individual country take-offs, so clearly indicated in the graphs in this chapter, were staggered and hence are masked when aggregated together.

Nevertheless, the yield growth was not solely attributable to the new varieties. They were necessary but not sufficient alone for success. Their potential could only be realized if they were supplied with high quantities of fertilizer and provided with optimal supplies of water. As was soon apparent, the new varieties yielded better than the traditional at any level of fertilizer application, although without fertilizer they sometimes did worse on poor soils [37]. Not surprisingly, average rates of application of nitrogen fertilizers, mostly ammonium sulphate and urea, doubled and redoubled over a very short period (Figure 4.9).

Because the new varieties were more exacting in their requirements, good irrigation, by providing a controlled environment for growth, became crucial. Most developing countries have both a dry and a wet growing season. Potentially, dry-season yields are 50–100 per cent greater than in the wet, but the lack of rainfall in the dry season and the high evapotranspiration rates resulting from absence of cloud cover make the crops liable to water stress. Without adequate irrigation, yields tend to be low and variable, whatever the level of fertilizer application. With irrigation and heavy fertilizer application, some of the highest cereal yields in the world have been attained. In South and South-East Asia the irrigated area grew from some 40 million hectares to over 65 million hectares between 1960 and 1980, a growth rate of 2·2 per cent per annum. By 1980 one-third of the rice area was irrigated, producing a rapid increase in the dry-season rice crop.

Just how much of the increased cereal production has been due to the availability of the new varieties, how much to increased fertilizer use and how much to the growth of irrigation is a matter of argument. An analysis of the eight countries (Bangladesh, Burma, China, India, Indonesia, Philippines, Sri Lanka and Thailand) responsible for 85 per cent of the

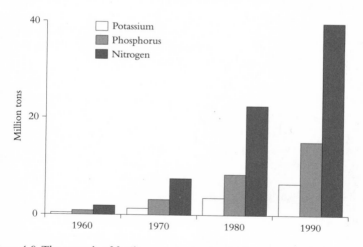

Figure 4.9 The growth of fertilizer use in the developing countries [38]

Asian rice crop suggested the three factors had a roughly equal contribution. Of the extra 117 million tons produced between 1965 and 1980, 27 million tons were attributable to the new varieties, 29 million tons to increased fertilizer use and 34 million tons to irrigation [39].

Inevitably, the need to provide high levels of fertilizer and controlled, irrigated environments meant that some locations were favoured over others. Uptake was fastest and most dramatic in those lands such as Sonora in Mexico, Luzon in the Philippines, the lowlands of Java, and the Punjab of India and Pakistan where irrigation was already well developed and where farmers, often larger than in the rest of the country, had good access to credit, had a greater propensity to take risks and were likely to be rapid adopters. Most important, innovation and adoption there was supported by governments willing and able to make and direct the necessary investments, including the necessary research structure which could take and adapt the new varieties to local conditions. These lands, scattered through Asia and Latin America, became the so-called Green Revolution lands.

In summary, the Green Revolution succeeded because it focused on three interrelated actions:

– Breeding programmes for staple cereals to produce early-maturing, day-length-insensitive and high-yielding varieties;
– The organization and distribution of packages of high-pay-off inputs, such as fertilizers, pesticides and water regulation;
– Implementation of these technical innovations in the most favourable agroclimatic regions and for those classes of farmers with the best expectations of realizing the potential yields.

In many ways this was a triumph for technology, rather than science. The dwarfing genes had been known for decades in China and Japan and most of the breeding techniques were well established. What made the difference was the investment, in China and independently in the rest of the developing world, in institutions and in the organization of delivery of the inputs necessary to make the science productive.

The success is undisputed, but it has been a revolution with serious limitations. In particular:

– Its impact on the poor has been less than expected;
– It has not reduced, and in some cases it has encouraged, natural-resource degradation and environmental problems;

– Its geographic impact has been localized; and
– There are signs of diminishing returns.

These are the issues I discuss in more detail in the following three chapters.

Notes

[1] H. Hanson, N. E. Borlaug and R. G. Anderson, 1982, *Wheat in the Third World*, Boulder, Colo., Westview Press

[2] The basis for determining who is 'hungry' has changed over time, and the figure would probably be lower if the current method of determining 'chronic undernourishment' was used. See D. Grigg, 1993, *The World Food Problem* (2nd edn), Oxford, Blackwell

[3] N. Alexandratos (ed.), 1995, *World Agriculture: Towards 2010. An FAO Study*, Chichester (UK), Wiley & Sons

[4] C. T. de Wit, H. H. van Laar and H. van Keulen, 1979, 'Physiological potential of food production', in J. Sneep and A. J. T. Henrickson (ed.), *Plant Breeding Perspectives*, Wageningen (Netherlands), Centre for Agricultural Publishing and Documentation (Publ. No. 118, PUDOC), pp. 47–82; D. L. Plucknett, 1993, *Science and Agricultural Transformation*, Washington DC, International Food Policy Research Institute (Lecture Series)

[5] Y. Ishizuka, 1969, 'Engineering for higher yields', in J. D. Eastin, F. A. Haskins, C. Y. Sullivan and C. H. M. van Bavel (eds.), *Physiological Aspects of Crop Yield*, Madison, Wis., American Society of Agronomy and Crop Science Society of America, pp. 15–26; Grigg, op. cit.

[6] R. F. Chandler, 1982, *An Adventure in Applied Science: A History of the International Rice Research Institute*, Los Banos (Philippines), International Rice Research Institute

[7] E. C. Stakman, R. Bradfield and P. C. Mangelsdorf, 1967, *Campaigns against Hunger*, Cambridge, Mass., Harvard University Press; C. de Alcantara Hewitt, 1976, *Modernizing Mexican Agriculture: Socioeconomic Implications of Technological Change, 1940–1970*, Geneva (Switzerland), UNRISD (Report No. 76.5)

[8] W. J. Lawrence, 1968, *Plant Breeding*, London, Edward Arnold (Studies in Biology No. 12)

[9] Stakman *et al.*, op. cit.

[10] ibid.

[11] Hanson *et al.*, op. cit.

[12] ibid.

[13] R. Barker, R. W. Herdt and B. Rose, 1985, *The Rice Economy of Asia*, Washington DC, Resources for the Future

[14] B. C. Wright, 1972, 'Critical requirements of new dwarf wheat for maximum production', in *Proceedings of the Second FAO/Rockefeller Foundation International*

Seminar on Wheat Improvement and Production, March 1968, Beirut, Ford Foundation

[15] The full list of the International Agricultural Research Centres is given in the Appendix; W. C. Baum, 1986, *Partners against Hunger: The Consultative Group on International Agricultural Research*, Washington DC, World Bank

[16] Chandler, op. cit.

[17] Barker *et al.*, op. cit.

[18] Stakman *et al.*, op. cit.; Barker *et al.*, op. cit.

[19] L. Brown, 1970, *Seeds of Change*, New York, NY, Praeger

[20] ibid.

[21] I. Hauri, 1974, *Le Projet Céréalier en Tunisie: Études aux Niveaux National et Local*, Geneva (Switzerland), United Nations Research Institute for Social Development (Report No. 74.4)

[22] Brown, op. cit.

[23] Grigg, op. cit.

[24] Plucknett, op. cit.

[25] ibid.

[26] Grigg, op. cit.

[27] ibid.

[28] D. Dalrymple, 1986, *Development and Spread of High Yielding Wheat Varieties in Developing Countries*, Washington DC, United States Agency for International Development; 1986, *Development and Spread of High Yielding Rice Varieties in Developing Countries*, Washington DC, United States Agency for International Development

[29] Hanson *et al.*, op. cit.

[30] M. Lipton and R. Longhurst, 1989, *New Seeds and Poor People*, London, Unwin Hyman

[31] Hanson *et al.*, op. cit.

[32] A. Pearse, 1980, *Seeds of Plenty, Seeds of Want: Social and Economic Implications of the Green Revolution*, Oxford, Clarendon Press; G. T. Castillo, 1975, *All in a Grain of Rice*, Laguna (Philippines), Southeast Asian Regional Center for Graduate Study and Research in Agriculture

[33] Barker *et al.*, op. cit.

[34] R. Chambers, 1988, *Managing Canal Irrigation: Practical Analysis from South Asia*, New Delhi (India), Oxford and IBH Publishing Co.

[35] R. F. Smith, 1972, 'The impact of the Green Revolution on plant protection in tropical and subtropical areas', *Bulletin of the Entomological Society of America*, 18, 7–14

[36] T. Dyson, 1996, *Population and Food: Global Trends and Future Prospects*, London, Routledge

[37] R. Barker, 1978, 'Yield and fertiliser input', in IRRI, *Changes in Rice Farming in Selected Areas of Asia*, Los Banos (Philippines), International Rice Research Institute

[38] FAO, 1994 [Fertilizer data diskettes], Rome, Food and Agriculture Organization

[39] R. W. Herdt and C. Capule, 1983, *Adoption, Spread, and Production Impact of Modern Rice Varieties in Asia*, Los Banos (Philippines), International Rice Research Institute; P. Pinstrup-Anderson and P. B. R. Hazell, 1985, 'The impact of the Green Revolution and prospects for the future', *Food Reviews International*, 1, 1–25

5 Food Production and the Poor

Green, yes; revolution, no . . . the achievement has not sufficed to improve poor people's food intakes much.

– Michael Lipton and Richard Longhurst,
New Seeds and Poor People [1]

The most controversial effect of the Green Revolution has been its impact on the lives of the poor [2]. In the opinion of many commentators, the new technologies were inherently biased in favour of the rich, benefiting large farmers at the expense of small, and landlords at the expense of tenants. They saw little evidence of the benefits 'trickling down' from the rich to the poor and argued that, without appropriate institutional reforms, efforts to introduce the new technologies were a waste of money. According to others, including Michael Lipton and Richard Longhurst, the technologies, despite their evident drawbacks, were essential if production was to be increased and labour demand maintained in the face of rapid population growth and little room for expanded cultivation. They posed the question: 'In the absence of the Green Revolution, would not the situation have been a great deal worse?' The answer to this question appears to be yes. Over the past thirty years world grain prices have declined in real terms and per capita calorie supplies in the developing countries have steadily increased. On average, everyone should be better fed and the incidence of poverty reduced.

Overall in the developing world, there has been a significant decline in the incidence of poverty over the past thirty years. Consumption per person increased by 70 per cent between 1965 and 1985, but this figure conceals a considerable variation [3]. Although there was growth almost everywhere in the 1960s, the regions began to diverge sharply in the 1970s, with little or no growth in Sub-Saharan Africa and only modest growth in South Asia (Figure 5.1). Several countries in East Asia have been particularly successful in reducing poverty, Indonesia taking less than a generation for the incidence of poverty to go from 60 per cent to under 20 per cent. Elsewhere progress has been modest. In India, poverty remained about 55 per cent from 1960 to 1974, fell to under 40 per cent by 1989, rose again to nearly 50 per cent between 1989 and 1992 and is now back to about the 1989 level [4]. Pakistan sharply reduced poverty in the 1970s, although there was less progress in the 1980s. Poverty in parts of Latin America increased in the 1980s but has begun to decline in the 1990s. For most of Sub-Saharan Africa poverty has continued to increase.

In this Chapter I investigate how far the Green Revolution has been successful in reducing hunger and alleviating poverty by asking three questions:

− Has producing cheaper and more widely available grain reduced under-nutrition and malnutrition?
− Has the Green Revolution benefited poor farmers, increasing their household incomes and consumption through greater production?
− Has there been a positive effect, direct or indirect, on the employment and incomes of rural labour?

The answers are complex and, inevitably, depend on geographic, social and political circumstances [5].

The impact of the Green Revolution on hunger has been uneven (Figure 5.2). Among the urban poor the incidence and severity of under-nutrition have declined, particularly in China, and also among the rural poor who live in the Green Revolution lands of East and South Asia, West Asia/North Africa, and Latin America, but probably not elsewhere. At best, undernutrition in the non-Green Revolution lands has been prevented from growing; at worst, many poor farm households in these lands may be getting less food, since their grain yields have increased little and they are receiving lower prices [6]. In Sub-Saharan Africa both the proportion and numbers of undernourished have risen.

The urban poor have benefited most wherever government policies have ensured that the extra grain has brought down domestic prices. In Colombia the growth in rice supplies based on cultivation of the new rice varieties has restrained food prices, with real benefits to the poor, estimated as equivalent to an over 12 per cent increase in income for households with fewer than 6,000 Colombian dollars per annum [7]. The Indian government, in contrast, has preferred to use the increased domestic

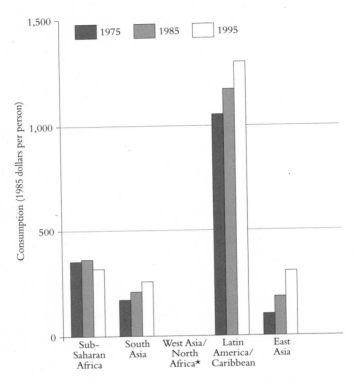

Figure 5.1 Consumption per person in the developing world [3]
★Figures are not available for West Asia/North Africa

production to replace imports and to build stocks [9]. Potential famines, as would have occurred following the 1987 drought, have been averted, but the overall impact on chronic undernutrition has been small.

Some of the most visible and significant improvements in nutrition have come from using the increased availability of food supplies to support highly targeted interventions [10]. Targets may be chosen in different ways. Subsidized coarse-grain prices have benefited the poor in Egypt. Food stamps have been provided in Jamaica, through registered health

Figure 5.2 Changes in numbers and proportion (percentages) of chronically under-nourished in developing countries [11]

centres, to well-defined groups such as pregnant and lactating women and children under five. Another approach has been to target certain disadvantaged regions, as in the Philippines. There, subsidized rice and cooking oil have been made available to selected poor villages through local retailers. The Tamil Nadu Integrated Nutrition Project in southern India has combined a number of targets, focusing on children six to thirty-six months of age in six districts with the lowest calorie consumption in the state. The food supplementation is provided only until the children achieve a satisfactory weight gain, when it ceases, so avoiding long-term dependence.

The next question is: how have the producers fared? Most poor farms are small – usually less than a hectare of arable land – and, not surprisingly, poor farmers were initially reluctant to take up the new varieties [12]. They were often tenants or sharecroppers, with poor security. And they were predominantly subsistence farmers, with no surpluses to be sold and hence no income to invest. Credit and inputs, such as fertilizers, were relatively more expensive for them because of the extra costs of packaging and servicing them in small amounts. Taking up the new technologies appeared to be a costly and highly risky undertaking, despite the promise of much higher yields. However, this situation soon began to change. Cheap credit became available, inputs were subsidized, and landlords, experiencing the benefits on their own land, encouraged their tenants to follow suit. Small farmers began to innovate, albeit step by step and at a slower rate than large farmers [13]. F. Bari, of the Bangladesh Academy for Rural Development, describes a farmer who took four years to fully adopt the new technologies; in the first year he opposed the idea on religious grounds, but in the second he planted the new seed and started to use irrigation water, and in the fourth year he applied fertilizer [14]. Soon, throughout the Green Revolution lands, virtually all farmers had become engaged in the process of change, the more conservative following the more adventurous as the benefits became clearly apparent.

By the mid-1970s small farmers had caught up with large farmers and, in several situations, were the first adopters [15]. Sometimes, they did not stay with the new technologies, reverting to the traditional varieties because of crop failures or lack of extension and other support. Problems also occurred if farmers diverted too much attention from other activities to the new varieties. They could lose their access to credit and risk serious consequences in the event of failure. But, in general, although large farmers were the biggest beneficiaries, small farmers in the Green Revolution lands soon began to see their productivity and incomes rise, and in amounts that were significant.

Once committed, small farmers often put a higher proportion of their land into the new varieties than did large. They also tended to lead in the use of fertilizers and insecticides while, not surprisingly, large farmers were the early adopters of tractors and mechanical threshers [16]. Both small and large farms can call on family labour, but small farms, because of their size, tend to use labour more intensively, and can benefit from the returns to labour-intensive crops [17]. Traditionally, this has meant better weed control, but it has also allowed small farms to get more return from modern inputs, such as fertilizers and irrigation water, where there is a complementary labour demand. By contrast, they are less able to benefit from the introduction of farm machinery. It is expensive and, although often available for hire, the cost per hectare is likely to be high because of the small size of the operation. Availability may also be constrained, renters having to wait until the owners of the machinery have finished the operations on their own land, causing ploughing or harvesting to be delayed with consequent loss of yield. For larger farmers, family labour is unlikely to be enough. They then have a choice of hiring labour or installing machinery. Increasingly they have chosen the latter option and, as I discuss below, have thereby captured a greater share of the returns from the new varieties.

Large farmers also tended to gain the so-called 'innovators' rent' which goes to those who first adopt a new technology [18]. As production increased, prices fell and later adopters received a more modest return. In some countries, large farmers became wealthy very quickly, using their new income to buy up land from the poor. Extended irrigation and higher yields in Mexico's Green Revolution lands brought great prosperity but it was increasingly concentrated in the hands of a small number of farmers, estimated at less than 200, each with an average of 500 hectares of land [19].

However, farm size was only one factor in determining uptake of the new technologies. Usually, the topography of the land, the quality of the soil and, most critical, the access to irrigation have been more important than the amount of land available [20]. The adopters had slightly less land than the non-adopters in Mexico, but their land was of better quality [21]. In India adoption was strongly correlated with water supply. Where irrigation was well developed, for example in the Punjab and Haryana, adoption was 100 per cent for all farm sizes, but in the eastern states of India, where water control is poor, adoption was low, under 50 per cent, even on irrigated land. The nature of the irrigation system was important; tube-wells usually provide a better controlled source of water than canal irrigation for the new varieties although, because tube wells are expensive investments, they tend not to be affordable by small farmers. In India, and elsewhere in the developing world, the new varieties have had less impact on the less favoured lands – unirrigated, with problem soils and topographies. It has only been in recent years, with the development of new rain-fed rices and high-yielding sorghums and millets, that the Green Revolution has moved outside the Green Revolution lands. For the most part, small farmers on less-well-favoured lands have received few benefits and in some cases have become poorer, at least as producers, as the price of grain has fallen.

The size, composition and livelihood strategy of the farm household are also important in determining whether farmers innovate. Households make complex decisions which take account not only of the likely benefits of adoption of new technologies, but also their opportunity costs (see Chapter 9). Success has been greatest where households have a tradition of being 'outward looking'. Innovators tend to have more 'experience of the world', gained by members of the household working full- or part-time in nearby towns. Off-farm income, commonly as high as one-third of net household income, is often crucial, providing cash to purchase inputs and a basis for taking risks [22]. And a market for cash crops may be an important incentive. In East Asia, where rice is the main subsistence crop, the new varieties have enabled farmers, particularly medium-sized farmers, to maintain their level of rice production on less land, so freeing land for more profitable crops for sale [23].

The third question is: what has been the impact on employment and the incomes of rural labour? From the outset, the Green Revolution was perceived as providing opportunities for, yet also posing a threat to, the livelihoods of agricultural workers. In some countries, the initial capture of the benefits by large farmers produced a scramble for jobs. In December of 1968 in Tanjore, one of India's model agricultural-development projects, forty-two persons were burned to death in a tragic clash between two groups of landless labourers. They were fighting over how best to get a share of the benefits from the new varieties being planted by the landowners in the district. One group of landless was willing to work at prevailing wage rates, while the others wanted to enforce a boycott until landlords were willing to raise the wages and share some of the new wealth [24].

Many of the characteristics of the new varieties and their associated technologies have the potential to increase rural employment. Extra labour is required to meet the higher standards and densities of sowing and planting (for example in the preparation of seed beds for the new rices), to ensure precise water control, to weed, to apply the greater amounts of fertilizer and pesticide, and to gather and thresh the larger harvests. But the biggest impact has come from the widespread growth in double- and triple-cropping on irrigated land, made possible by the shorter growing time of the new varieties. In India it was soon realized that because the new wheats could be planted in December, thirty days later than the traditional wheats, and still be harvested at the usual time in the spring, it was possible to insert a rice crop beforehand. George Blyn of Rutgers University describes the pattern of cropping on a farm in the Indian state of Haryana [25]. The owner, Atwahl Swaran Singh, has eleven hectares on which he grows four crops in the year: a new wheat harvested in the spring; followed by a green manure; then a short-duration rice or potatoes; leading up to the main wheat planting. Large areas of land stretching from the Punjab to West Bengal have adopted similar rice–wheat rotations [26].

Continuous rice-cropping is also possible in some locations, and in the Philippines one ingenious farmer developed a rice garden in which rice was being planted every few days, creating an intensive mosaic of cultivation.

72

Farmers in the Chiang Mai valley of northern Thailand, which contains some 150,000 hectares of arable land (see Chapter 10), have developed over twenty different systems of crop rotation involving traditional and new rices, pulses and vegetables or other cash crops, such as tobacco and soybeans. In these situations the year round demand for labour has increased and stabilized, reducing the fluctuations in employment and incomes (Figure 5.3).

But, countering these trends, the Green Revolution has provided strong incentives for mechanization [27]. Increased incomes, especially on larger farms, have produced capital for investment but, more important, have increased the power of large farmers to lobby for substantial credit and subsidies. In Sri Lanka the incentives have included reduced rates of duty on tractors, preferential foreign-exchange allocations, tax concessions on depreciation allowances, and loans which are characterized by lower commercial criteria, subsidized rates of interest and restrictions on repossession in the event of default [28]. Real interest rates on loans to purchase tractors in Brazil in the 1960s were negative, ranging from −4 to −42 per cent [29]. And the consequent cost savings for farmers have been considerable. It cost 495 rupees, in India in the 1960s, to pump 10 acre-inches of water by hand, 345 rupees if pumped by draught animals powering a Persian wheel and only 60 rupees using a diesel pump [30].

The pace of mechanization has varied from country to country. China has pursued a policy of rapid mechanization, initially as a means of promoting

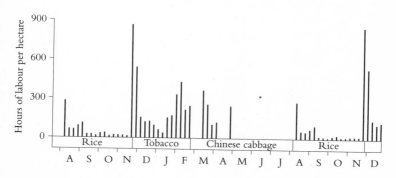

Figure 5.3 Labour demands for triple-cropping in the Chiang Mai valley, Thailand [31]

collectivization. Between 1965 and 1980 the number of tractors grew at 17 per cent per year and for two-wheeled power tillers the annual growth rate was over 50 per cent. This slowed after the shift to household-based production at the beginning of the 1980s. There was a sharp reduction in tractor cultivation and an even greater reliance on power tillers producing, as in other market economies, a variation in the mixes of sources of power – tillers, animal and human labour – on individual farms [32]. Power tillers have also become common in wetland areas of South-East Asia, although their adoption has been slower in the rest of Asia and other regions of the developing world. The commonest machines in both China and India in the 1980s were pumps for irrigation and drainage. Tractor use has also grown rapidly in West Asia/North Africa, but has been slower and less pervasive in Latin America. Least mechanized is Sub-Saharan Africa (Figure 5.4). Although, in theory, mechanization should be most common where labour is scarce and land abundant, in practice the degree of agricultural intensification has been more decisive [33].

These various forms of mechanization have very different consequences for production and employment. Using machines to spray insecticides can increase yields and, generally, does not displace labour. Mechanical irrigation devices, such as water pumps, produce higher yields, and may or may not be labour-displacing. By contrast, the spraying of herbicides and the introduction of tractors and two-wheeled power tillers to prepare land, and of machines to harvest, thresh and mill grain, produce relatively little direct effect on yields, and are strongly labour-displacing [34].

The biggest impact comes from the introduction of combine harvesters.

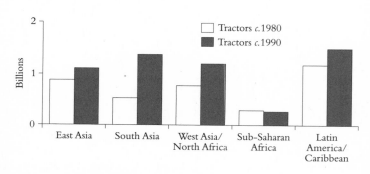

Figure 5.4 Four-wheel tractors in the developing world [35]

An analysis by Gerald Gill of Winrock International of the introduction of combines into the Chilalo region of Ethiopia (see page 79 below) suggests that a modest combine can harvest a hectare of wheat in an hour to an hour and a half, compared with as many as 50 person-days using sickles, ox treading and winnowing [36]. In a study of the Punjab and neighbouring states of northern India, H. Laxminarayan and his colleagues at the University of Delhi estimated that combines led to a 95 per cent reduction in employment, hitting especially hard the seasonal migrants who come from the poorer states to the east [37]. According to their calculations, if all farmers in the Punjab cultivating more than four hectares were to adopt combines there would be a loss of over 40 million person-days of labour, and without any evident increase in farm production or in cropping intensity. The sole advantage would be to relieve the farmer's burden of recruitment and supervision of large gangs of casual labour.

The effects of mechanization have been most pronounced in rice cultivation. This is inherently more labour-intensive than the cultivation of the other main staple cereals, requiring up to double the labour requirement per hectare [38]. It is one of the reasons why rice farms tend to be small, usually under two hectares, and heavily reliant on family labour. A rice-farming family in Java will put in an average of over forty hours a week throughout the year. Women's labour is particularly important, contributing between one-third and one-half of the total labour in rice-farming systems [39]. In South and East Asia, women traditionally have done most of the transplanting, weeding and harvesting, although there are regional variations. As a result, in some countries there has been greater employment of women, in particular from landless households. In the early years of the Green Revolution in India, hired female labour was higher on farms cultivating the new varieties, largely because of the increased demand for weeding and harvesting [40]. But the effect has not been universal. There have been declines in both female and male labour demand in the Philippines and Indonesia, resulting from mechanical direct seeding, the introduction of multi-row transplanters, the use of rotary weeders and, most important, new harvesting practices and mechanized threshing and milling [41].

The traditional method of harvesting rice in Indonesia was to use a small *ani-ani* knife. The work was usually performed by women, who cut the rice stalks about six inches below the panicles of grain. Each harvester received from the farmer a proportion of the grain she harvested [42]. The share was one-quarter to the farmer's relatives, between one-quarter and one-sixth to close neighbours and one-tenth to fellow villagers. It was a somewhat wasteful system, since the large numbers of workers caused significant losses from trampling, stealing and handling. But because harvesting was open to all members of the community, the system provided a guarantee of income even for the poorest households. The *ani-ani* began to be replaced by the sickle in the 1950s, and with the introduction of the new varieties which mature almost simultaneously there was pressure for a quicker, better-organized harvest. In the 1970s, men took over the harvesting, initially for a fixed or somewhat lower proportion of the harvest on the traditional basis, or in return for weeding and other work. Eventually a contract system was introduced, largely as a response to the growing numbers of landless labourers that were making it difficult for farmers to control the number of harvesters and their share. The standing crop was sold to middlemen ten days before harvest; they hired workers who were paid in cash. Similar contracts have been introduced for weeding.

One of the most significant impacts on labour demand in Asia has been the introduction of mechanical hullers to replace the hand-milling of rice. In Bangladesh, rice is traditionally milled using a foot-operated pestle and mortar, called a *dheki* [43]. It provides much-needed employment, particularly for poor landless women, often being the only source of income for divorcees and widows, in a culture where women need to observe the seclusion of purdah. A mill reduces the labour input from 270 to 5 hours per ton. By the 1980s over 40 per cent of rice in Bangladesh was being mechanically milled, and the annual introduction of 700 new mills was replacing 100,000 to 140,000 women a year.

The most powerful argument for the introduction of tractors and other machines is the possibility of moving to higher cropping intensity, which, in turn, should lead to increased overall employment. But various studies suggest the losses outweigh the gains. The higher cropping intensity coupled with related technology of wheat−rice systems in the western part of the Indian state of Uttar Pradesh has increased the labour demand by over 60 hours per hectare, but the introduction of tractors results in a loss of over 110 hours [44].

Even where there have been substantial gains they have not lasted. In Taiwan, where the agricultural transformation happened earlier than in most developing countries, labour demand consequent on multiple-cropping reached a peak in the 1960s, but then began to decline with increasing mechanization [45]. In Figure 5.5, Burma in the 1930s represents the pre-Green Revolution pattern of low yields produced by a low input of human labour supplemented by animal power. By the 1970s the new varieties were being planted in countries such as Sri Lanka. Fertilizers and pesticides were being applied; labour use was high but yields had yet to reach their potential. The contrast between Taiwan in the 1960s and 1970s shows the effects of mechanization. Yields of 4–5 tons were achieved, initially with high labour input but then falling back, not to pre-Green Revolution levels, but by at least a third from the maximum attained. A similar pattern is emerging in the Green Revolution lands. In Laguna, in the Philippines, wet-season labour per hectare rose from 86 to 112 days between 1966 and 1975 but then, owing to the growing use of threshers and herbicides, fell back to 93 days in 1981. The proportion of hired labour rose from 60 to 80 per cent and then reduced to just over 70 per cent [46].

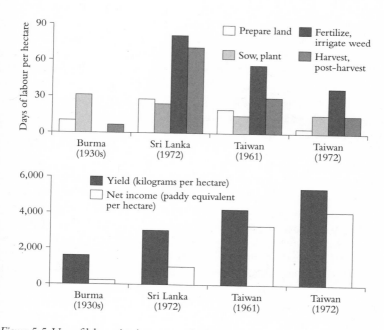

Figure 5.5 Use of labour in rice systems in Asia [47]

In the early years of the Green Revolution, the introduction of irrigation, the increased use of fertilizers and modern varieties raised labour demand by about 20 per cent per hectare [48]. Doubling of yields increased demand by as much as 30–50 per cent [49]. But the subsequent growth of mechanization has partly reversed these gains. Doubling of yields today produces only a 10–30 per cent increase [50]. At the same time, there has been a steady growth in rural labour supply throughout much of the developing world, of the order of 2–3 per cent per annum [51]. Both trends have restrained the growth in real wage rates.

Even where there has been a major employment impact of the Green Revolution, for example in the Indian Punjab, the rise in real wage rates has been very slow [52]. Seasonal immigrants from remote, poor areas of Bihar and eastern Uttar Pradesh have been prepared to work for near-subsistence wages. And whenever labour has become too expensive, the large landowners have been able to draw on credit and subsidies to mechanize further. There has been net emigration from the Punjab in recent years, with an increasing proportion of out-migrants coming from the scheduled and other low castes [53].

For the Green Revolution lands as a whole, real wage rates have remained the same, or risen only a little, and labour's share in farm income has fallen, because the price of land has grown relative to that of labour [54]. For example, in one district of South India the ratio of rent to wage doubled in the 1970s [55]. As Michael Lipton and Richard Longhurst put it, with the introduction of the new varieties, 'employment of labour goes up somewhat, the real wage does not go up a lot, and the rewards (price, rent) of land go up a good deal, probably reducing labour's share in income'. Effectively, the major benefits of the new varieties have gone to the landowners rather than to the labourers. One consequence is a widening gap between those who own or rent a little land and those who are truly landless. The number of landless has risen rapidly, although it is difficult to get good estimates because census definitions are not always consistent. For the landless the only option, if local employment is not available, is to migrate to other rural areas where wage rates are higher, or to the cities.

Underlying these effects there has been a change in traditional social and economic relationships [56]. A more subsistence-based, communal self-help system of agriculture has been replaced by one which relies entirely on market forces. The poor and landless who, in the past, could obtain food for work, or from labour-exchange schemes, or indeed gather food from communal or undeveloped land, no longer can do so. Holdings have become larger and the power of the landowners has grown. At the same time, many of the benefits have gone to outsiders.

Under traditional agriculture, only land, labour and limited capital, for example draught animals and land improvement, are used in production and the income is distributed to the owners of each of these factors. But as economies develop, there are new beneficiaries, increased income flowing to the providers of the new technology, to the urban or foreign suppliers of inputs, especially fertilizers, and to the producers and repairers of machines, often leaving both labour and land with a smaller share of gross revenue. Nevertheless, on the positive side of the ledger, farm workers get cheaper food so their wages buy more. And without the new varieties wage rates would have fallen further [57]. The question is whether, overall, equity has increased or decreased. According to most village studies, all classes benefit from the introduction of irrigation combined with the planting of the new varieties. But the crucial determinants are the extent of labour-displacing mechanization and the availability of alternative employment.

A classic example of growing inequity is provided by the Chilalo Agricultural Development Project in Ethiopia, analysed by John Cohen of Cornell University. The project, which was begun in 1967, was targeted on improving the livelihoods of some 600,000 households cultivating wheat, barley and flax. They were provided with low-cost credit to purchase fertilizers and seeds of new varieties. The water supply and feeder roads were improved and programmes of community education and development initiated. Yields and overall production grew, with real increases in income of 69–90 per cent among the participating households. But the benefits were quickly captured by the rural elites. By 1972 large landowners had introduced tractors and combines and were engaged in eviction of tenants. The tenancy rate was about 50 per cent prior to the project; by 1972 it had declined in some parts to below 12 per cent, with many thousands of tenants being thrown off the land. The prices of land rose, as did the rents, and much of the income gains to the tenants and small farmers went to pay corrupt officials. Because of the weakness of the local economy there were few employment opportunities for those displaced from the land.

By contrast, the new technologies, when placed within a broader context of agricultural development, can provide an equitable spread of benefits.

An illuminating example is the recent history of North Arcot in southern India, studied by Peter Hazell of IFPRI and C. Ramasamy of Tamil Nadu Agricultural College [58]. North Arcot is a rice-growing district, dependent on tank and tube-well irrigation. It is not a prime Green Revolution area and the overall benefits have not been as great as in the Punjab, but have been significant: rice output grew by nearly 60 per cent between mid-1960s and late 1970s. Most of the farms are small (average 1·2 hectares) and one-third of rural households consist of landless labourers. Initially the larger farmers (about 2 hectares in size) were the adopters. However, by the early 1980s over 90 per cent of the land was planted to new varieties, with no difference between large and small farms, largely as a result of improved access to credit and the release of new varieties better adapted to small firms with less reliable water.

The installation of irrigation pumping and threshing machines produced a fall in labour demand, family labour increasing while hired labour fell by 25 per cent per farm. This would have been worse if there had been an increase in use of tractors, which was probably inhibited by the small farm size. Nevertheless, wage rates rose as did employment earnings, for two reasons:

– Partly as a consequence of the growing rice economy, there were increased opportunities in dairying and in non-farm activities: government employment programmes, a growing silk-weaving industry and numerous small-scale service and household industries (each dollar of income generated in agriculture generated another 80 cents in non-farm activity).
– There was substantial migration to urban areas.

As a result there was no increase in landlessness and little or no increase in size of farms. All groups benefited from a growth in income and consumption expenditure and in their diets: daily calories per adult equivalent rose, on average, among rice farmers from about 1,800 calories to over 3,000 and for landless labourers from 1,700 to over 2,500 calories (Figure 5.6). The relatively poorer benefits to large farms were due to sharp increases in costs of fertilizers and labour.

In summary, there has been a significant reduction in poverty and hunger in those countries directly affected by the Green Revolution. The urban poor have had the benefit of cheaper food. Small farmers have benefited from higher incomes in the Green Revolution lands and so have the landless with, in some situations, rising real wages. But the introduction

80

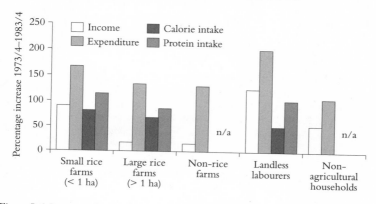

Figure 5.6 Increases in standard of living in North Arcot, southern India [59]

of mechanization has tended to erode these benefits. And in many of the rural areas largely untouched by the Green Revolution both poor farmers and the landless have suffered losses in real income and increased hunger. As Michael Lipton and Richard Longhurst rightly said in the quote at the beginning of this chapter, the benefits to the really poor have not been that much. The core of the Green Revolution technology – the new varieties – are potentially poverty-alleviating. Higher yields, producing cheaper food, and the greater labour demand consequent on multiple-cropping can reduce poverty and hunger, but only if technological innovation is positioned within a broad-based agricultural and rural development. Technologies are by themselves not enough. As Michael Lipton puts it, 'there is a limit to technical cures for social pathologies' [60]. Too often the new technologies have been injected into communities with rapidly growing populations already dominated by excessive inequalities where, in the absence of countervailing policies, the powerful and the better-off have acquired the major share of the benefits. As a consequence a high proportion, over 20 per cent, of the developing world's population is still poor and hungry.

Notes

[1] M. Lipton and R. Longhurst, 1989, *New Seeds and Poor People*, London, Unwin Hyman

[2] R. Barker, R. W. Herdt and B. Rose, 1985, *The Rice Economy of Asia*, Washington DC, Resources for the Future; Lipton and Longhurst, op. cit.; R. S. Anderson, P. R. Brass, E. Levy and B. M. Morris (eds.), 1982, *Science, Politics and the Agricultural Revolution in Asia*. Boulder, Colo., Westview Press; A. Pearse, 1980, *Seeds of Plenty, Seeds of Want: Social and Economic Implications of the Green Revolution*, Oxford, Clarendon Press; V. W. Ruttan and H. P. Binswanger, 1978, 'Induced innovation and the Green Revolution', in H. P. Binswanger and V. Ruttan (eds.), *Induced Innovation: Technology, Institutions and Development*, Baltimore, Md, Johns Hopkins University Press, pp. 358–408; Y. Hayami and M. Kikuchi, 1981, *Asian Village Economy at the Crossroads*, Tokyo (Japan), Tokyo University Press

[3] World Bank, 1990, *World Development Report, 1990*, London, Oxford University Press

[4] G. Datt and M. Ravallion, 1996, 'Macroeconomic crises and poverty monitoring: a case study for India', Washington DC, Food Consumption and Nutrition Division, International Food Policy Research Institute (Discussion Paper No. 20)

[5] Lipton and Longhurst, op. cit.

[6] ibid.

[7] G. M. Scobie and R. Posada, 1978, 'The impact of technical change on income distribution: the case of rice in Colombia', *American Journal of Agricultural Economics*, 60, 85–92

[8] World Bank, 1998, *World Development Indicators 1998*, Washington DC, World Bank

[9] Lipton and Longhurst, op. cit.

[10] N. Alexandratos (ed.), 1995, *World Agriculture: Towards 2010. An FAO Study*, Chichester (UK), Wiley and Sons

[11] FAO, 1992, *World Food Supplies and Prevalence of Chronic Undernutrition in Developing Regions as Assessed in 1992*, Rome (Italy), Food and Agriculture Organization (Document ESS/MISC/1992)

[12] F. R. Franke, 1971, *India's Green Revolution: Economic Gains and Political Costs*, Princeton, NJ, Princeton University Press

[13] B. Lockwood, P. K. Mukherjee and R. T. Shand, 1971, *The HYV Program in India*, Part I, New Delhi (India), Planning Commission (India) with Australian National University; M. G. G. Schluter, 1971, *Differential Rates of Adoption of the New Seed Varieties in India: The Problem of Small Farms*, Ithaca, NY, USAID/Department of Agricultural Economics, Cornell University (Occasional Paper No. 47)

[14] F. Bari, 1974, *An Innovator in a Traditional Environment*, Comilla (Bangladesh), Bangladesh Academy for Rural Development

[15] M. Prahladchar, 1983, 'Income distribution effects of the green revolution in India: a review of empirical evidence', *World Development*, 11, 927–44; B. Farmer (ed.), 1977, *Green Revolution?* London, Macmillan

[16] R. Barker and R. Herdt, 1978, 'Equity implications of technology changes', in IRRI, *Interpretive Analysis of Selected Papers in Rice Farming in Selected Areas of Asia*, Los Banos (Philippines), International Rice Research Institute

[17] H. P. Binswanger, K. Deininger and G. Feder, 1995, 'Power, distortions, revolt and reform in agricultural land relations', in J. Behrman and T. N. Srinivasan, *Handbook of Development Economics*, Vol. IIIB, Amsterdam (Netherlands), Elsevier, pp. 2661–772

[18] H. Binswanger, 1980, 'Income distribution effects of technical change – some analytical issues', *South East Asian Economic Review*, 1, 179–218; D. Dalrymple, 1979, 'The adoption of high yielding grain varieties in developing countries', *Agricultural History*, 53, 704–26

[19] C. de Alcantara Hewitt, 1976, *Modernizing Mexican Agriculture: Socioeconomic Implications of Technological Change, 1940–1970*, Geneva (Switzerland) (UNRISD Report No. 76.5)

[20] Lipton and Longhurst, op. cit.

[21] R. V. Burke, 1979, 'Green Revolution technology and farm class in Mexico', *Economic Development and Cultural Change*, 28, 135–54

[22] F. Chuta and C. Liedholm, 1979, *Rural Non-farm Employment: A Review of the State of the Art*, East Lansing, Mich., Michigan State University

[23] F. Bray, 1986, *The Rice Economies: Technology and Development in Asian Societies*, Berkeley, Calif., University of California Press

[24] *New York Times*, 28 December 1968, p. 3, quoted in L. Brown, 1970, *Seeds of Change*, New York, NY, Praeger

[25] G. Blyn, 1983, 'The green revolution revisited', *Economic Development and Cultural Change*, 31, 705–27

[26] H. Hanson, N. E. Borlaug and R. G. Anderson, 1982, *Wheat in the Third World*, Boulder, Colo., Westview Press

[27] H. Binswanger, 1978, *The Economics of Tractorization in South Asia*, Washington DC, Agricultural Development Council

[28] J. Farrington and F. Abeyratne, 1984, 'The impact of small farm mechanisation in Sri Lanka', in J. Farrington, F. Abeyratne and G. J. Gill, *Farm Power and Employment in Asia*, Colombo (Sri Lanka), Agrarian Research and Training Institute/Bangkok (Thailand), Agricultural Development Council

[29] W. R. Thirsk, 1985, *The Growth and Impact of Farm Mechanization in Latin America*, Washington DC, Agriculture and Rural Development Department, World Bank

[30] J. S. Balis, 1968, *An Analysis of Performance and Costs of Irrigation Pumps Utilizing Manual, Animal and Engine Power*, New Delhi (India), Agency for International Development

[31] P. Gypmantasiri, A. Wiboonpongse, B. Rerkasem, I. Craig, K. Rerkasem, L. Ganjanapan, M. Titayawan, M. Seetisarn, P. Thani, R. Jaisaard, S. Ongprasert, T. Radnachaless and G. R. Conway, 1980, *An Interdisciplinary Perspective of Cropping Systems in the Chiang Mai Valley: Key Questions for Research*, Faculty of Agriculture, University of Chiang Mai (Thailand)

[32] World Bank, 1987, *Agricultural Mechanization: Issues and Options*, Washington DC, World Bank

[33] P. Pingali, Y. Bigot and H. P. Binswanger, 1987, *Agricultural Mechanization and the Evolution of Farming Systems in Sub-Saharan Africa*, Baltimore, Md, Johns Hopkins University Press

[34] Barker *et al.*, op. cit.

[35] FAO, 1995, *FAO Yearbook, Production Vol. 48*, Rome (Italy), Food and Agriculture Organization (FAO Statistics Series No. 125)

[36] G. J. Gill, 1991, *Seasonality and Agriculture in the Developing World: A Problem of the Poor and Powerless*, Cambridge, Cambridge University Press

[37] H. Laxminarayan, D. P. Gupta, P. Rangaswami and R. P. S. Malik, 1981, *Impact of Harvest Combines on Labour Use, Crop Pattern and Productivity*, New Delhi (India), Agricultural Economics Research Centre, University of Delhi and Agricole Publishing Academy

[38] S. Ishikawa, 1978, *Labour Absorption in Asian Agriculture*, Bangkok (Thailand), International Labour Office

[39] Barker *et al.*, op. cit.; E. Boserup, 1970, *Women's Role in Economic Development*, New York, NY, St Martin's Press

[40] B. Agrawal, 1984, 'Rural women and high-yielding rice technology', *Economic and Political Weekly*, 19, A39–52

[41] A. Res, 1983, 'Changing labor allocation patterns of women in Iloilo rice farm households'; P. Sajogyo, 1983, 'Impact of new farming technology in women's employment'; B. White, 1983, 'Women and the modernization of rice agriculture: some general issues and a Javanese case study', papers presented at the Conference on Women in Rice Farming Systems, International Rice Research Institute, Los Banos (26–30 September 1983), Philippines

[42] W. Utami and J. Ihalauw, 1973, 'Some consequences of small farm size', *Bulletin of Indonesian Economic Studies*, 9, 46–56

[43] M. Cain *et al.*, 1979, *Class, Patriarchy and the Structure of Women's Work in Rural Bangladesh*, New York, NY, Center for Policy Studies, Population Council (Working Paper 43); M. Greeley, 1987, *Post-harvest Losses, Technology and Unemployment: the Case of Bangladesh*, Boulder, Colo., Westview; G. C. Scott and M. Carr, 1985, *The Impact of Technology Choice on Rural Women in Bangladesh: Problems and Opportunities*, Washington DC, World Bank (Staff Working Paper 731)

[44] P. K. Joshi, D. K. Bahl and D. Jha, 1981, 'Direct employment effect of technical change in UP agriculture', *Indian Journal of Agricultural Economics*, 36, 1–4

[45] T. Lee, H. Chan and Y. Chen, 1980, 'Labour absorption in Taiwan agriculture', in ILO, *Labour Absorption in Agriculture: The East Asian Experience*, Bangkok (Thailand), International Labour Organization, pp. 167–236

[46] M. Kikuchi and Y. Hayami, 1983, *New Rice Technology, Intra-rural Migration and Institutional Innovation in the Philippines*, Los Banos (Philippines), International Rice Research Institute (Research Paper No. 86)

[47] Barker *et al.*, op. cit.

[48] Barker and Herdt, op. cit.

[49] S. K. Jayasuriya and R. T. Shand, 1986, 'Technical change and labour absorption in Asian agriculture: some emerging trends', *World Development*, 14, 415–28; Lipton and Longhurst, op. cit.

[50] Lipton and Longhurst, op. cit.; D., V. K. and R. Singh, 1981, 'Changing patterns of labour absorption on agricultural farms in Eastern U.P.', *Indian Journal of Agricultural Economics*, 36, 39–44

[51] Lipton and Longhurst, op. cit.

[52] S. Bhalla, 1979, 'Real wage rates of agricultural labourers in the Punjab, 1961–1977: a preliminary analysis', *Economic and Political Weekly*, 14, A57–A68; Blyn, op. cit.; M. J. Leaf, 1983, 'The green revolution and cultural change in a Punjab village', *Economic Development and Cultural Change*, 31, 227–70.

[53] A. S. Oberai and H. K. Manmohan Singh, 1980, 'Migration flows in Punjab's Green Revolution belt', *Economic and Political Weekly*, 16, A2–4

[54] Lipton and Longhurst, op. cit.

[55] V. Rajagopalan and S. Varadarajan, 1983, 'Nature of new farm technology and its implications for factor shares – a case study in Tamil Nadu', *Indian Journal of Agricultural Economics*, 38, 4, October–December

[56] Barker *et al.*, op. cit.

[57] Hayami and Kikuchi, op. cit.

[58] P. B. R. Hazell and C. Ramasamy, 1991, *The Green Revolution Reconsidered: The Impact of High-Yielding Varieties in South India*, New Delhi, Oxford University Press

[59] ibid.

[60] M. Lipton, pers. comm.

6   Food Production and Pollution

> Increased agricultural production is also limited by pollution. Industry
> is often to blame, but agriculture may be both culprit and victim.
> – Gordon Conway, Uma Lele, Jim Peacock and Martin Piñeiro,
> *Sustainable Agriculture for a Food Secure World* [1]

In the 1960s, when the Green Revolution was beginning to make its
impact, little thought was given to environmental consequences. They
were deemed either insignificant or, at least, capable of being easily redressed
at a future date, once the main task of feeding the world was accomplished.
There was also a strongly held view, one still commonly voiced, that a
healthy, productive agriculture would necessarily benefit the environment.
Good agronomy was good environmental management. It is a point
with some force. Traditional agriculture is usually informed by ecological
wisdom. Modern technologies can be as environmentally sensitive, but
only if they are designed and used with the benefit of modern ecological
knowledge.

Too often, over the past thirty years, the technologies accompanying
the Green Revolution have turned out to have adverse environmental
effects. The heavy use of pesticides has caused severe problems. There is
growing human morbidity and mortality while, at the same time, pest
populations are becoming resistant and escaping from natural control. In
the intensively farmed lands of both the developed and developing countries, heavy fertilizer applications are producing nitrate levels in drinking
water that approach or exceed permitted levels, increasing the likelihood
of government restrictions on fertilizer use. The contribution of agriculture
to global pollution has grown, with potentially serious consequences. And,
as I shall discuss in Chapters 13 and 14, there has been a steady deterioration
in the natural-resource endowment of the developing countries. Land has
become degraded, forests and biodiversity lost, grazing land and fisheries
overexploited. In many respects, agriculture is both culprit and victim.

Pesticides not only cause or aggravate pest problems (see Chapter 11), they contaminate the environment and may have serious consequences for human health [2]. Their effects on wildlife are well documented. Organochlorine insecticides, such as DDT and dieldrin, which were in common use in the 1950s and 1960s, are highly persistent. In the developed countries they caused dramatic declines in birds of prey, such as the peregrine falcon, and had less visible, yet equally serious, consequences on a great variety of other wildlife. Similar effects must have occurred in the developing countries, although the evidence is not as complete. In Suriname, on the north-east coast of South America, intensive spraying of ricefields with endrin resulted in accumulation of residues in the food chains and kills of fish-eating egrets, herons and other birds of prey. There is much anecdotal evidence of a similar nature; for example, there are many reports of death to cattle and other livestock. Pesticides are also responsible for the eradication of fish, shrimps and crabs in ricefields – important sources of protein for the poor and landless [3]. In recent years, the organochlorines have largely been replaced by organophosphate and pyrethroid insecticides, which are not as persistent. Judging from experience in the developed countries, the harmful effects on the environment are less, but not insignificant.

The use of pesticides in the developing countries grew rapidly in the 1960s and 1970s reaching a total of over half a billion tons (in terms of active ingredients) in the mid-1980s, about a fifth of world consumption [4]. East Asia accounts for nearly 40 per cent of developing-country use, but there has also been rapid growth in South Asia (Figure 6.1). In India, where most of the pesticides are produced domestically, the treated area expanded from 6 million hectares in 1960 to over 80 million hectares in the mid-1980s [5]. Nevertheless, the rates of use are still well below those of the developed and developing countries of East Asia. About half of developing-country consumption is of insecticides – compounds responsible for the most serious health and environmental problems – and half the world's consumption of insecticides is accounted for by the developing countries. In contrast, the developed countries account for 90 per cent of herbicide use, compounds which, in general, are safer.

Although pesticide use is lower than in the developed countries, the relative impact on human health in the developing countries is probably greater. Partly, this is because of the high proportion of insecticide use. And, more important, lack of appropriate legislation, widespread ignorance

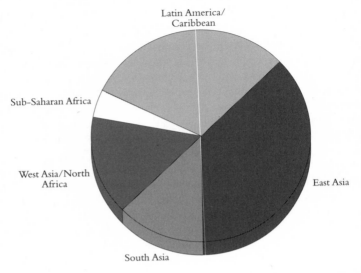

Figure 6.1 Share of pesticide use in the developing countries, in the mid-1980s [6]

of the risks involved, poor labelling, inadequate supervision and the discomfort of wearing full protective clothing in hot climates, greatly increase the risk of harm to both agricultural workers and the general public. The hazards are made worse by the continued use of compounds, such as DDT and chlordane, either banned or severely restricted in the developed countries. According to the US Food and Drug Administration, 5 per cent of foods imported into the USA in 1988 contained residues of banned pesticides, indicating their widespread use [7].

Reporting is often unreliable; the best estimates available suggest about half a million cases of accidental pesticide poisoning each year in the developing countries and 2,300 deaths (this compares with 360,000 cases in the developed countries and 1,000 deaths) [8]. But pesticide-related illnesses and deaths are likely to be seriously under-reported. The symptoms of pesticide poisoning are frequently confused with cardiovascular and respiratory diseases, or with epilepsy, brain tumours and strokes.

Central Luzon in the Philippines is one of the Green Revolution lands where pesticide spraying on rice has been especially intensive. Following the introduction of the new varieties, farmers were spraying four or five times in a season with organochlorine compounds classified as extremely or highly hazardous. In 1985 Michael Loevinsohn, a postgraduate student at Imperial College investigating the ecology of rice pests, became concerned when he saw how pesticides were being applied and decided to look at local mortality records. The results were startling. According to the records, over the period when pesticide use was doubling, death rates from diagnosed pesticide poisoning had risen by nearly 250 per cent, and by 40 per cent from other conditions which may have been related to pesticide use. The increased mortality was confined to men; for women and children it declined (Figure 6.2) [9]. Moreover, the deaths peaked in August when spraying is most intensive; after the introduction of double-cropping in the mid-1970s, a second peak appeared in February, which is when the newly cultivated dry-season crops are sprayed.

Loevinsohn's evidence is circumstantial, but it was powerful enough to

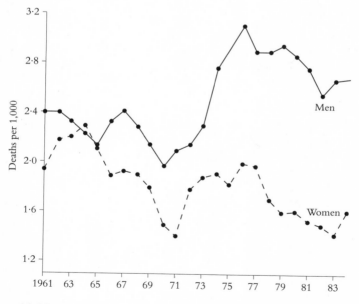

Figure 6.2 Non-traumatic deaths among rural men and women in Nueva Ecija, Luzon, the Philippines [10]

stimulate further investigation. Fifty per cent of rice farmers responding in a survey, conducted by Agnes Rola of the University of the Philippines, claimed sickness owing to pesticide use [11]. She and Praphu Pingali of IRRI followed this up with a detailed study which revealed a range of disease conditions in farmers exposed to pesticide use that were significantly higher than in those not exposed (Figure 6.3). These results seem to bear out Loevinsohn's findings and, together, they imply there are many thousands of deaths a year resulting from pesticide use in the Philippines. Although the modern organophosphate insecticides are much less toxic, there are growing fears that long-term exposure may result in damage to human nervous systems and the development of cancers.

High levels of organochlorine residues have accumulated in both rural and urban dwellers in the developing countries. The principal sources of contamination have been the spraying of mosquitoes for malaria control and the especially heavy insecticidal applications to non-food crops such as cotton, but residues have appeared in human foods. It is common in India for diets to contain pesticide levels greater than the Acceptable Daily Intake (ADI) (Figure 6.4) [13]. There are also reports from Malaysia of high residue levels in fish caught in ricefields.

The growth in fertilizer use in the developing countries has also brought problems in its wake, although fortunately some of the initial fears have proven unfounded. Ideally, all the nutrients in fertilizers are taken up by

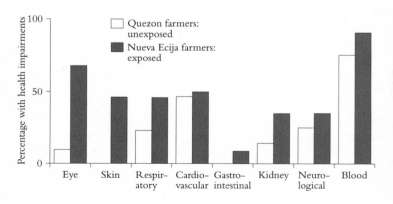

Figure 6.3 Health impairment of farmers exposed and unexposed to pesticide use in the Philippines [12]

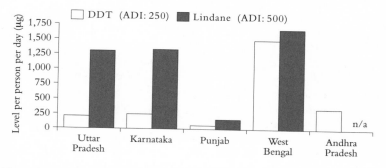

Figure 6.4 Levels of pesticide residues in Indian diets [14]. The ADI is the Acceptable Daily Intake

growing crops; in practice much is lost, up to 70 per cent in rice fields. In the seasonal tropics, nutrients build up in the dry season and then run off the fields with the onset of the monsoon. Losses are also high under irrigation. Application of nitrogen fertilizers, whether synthetic or organic (derived from plant wastes or animal manures), thus results in growing levels of nitrates in run-off waters and, through percolation, in subsoil water and aquifers. In the Green Revolution lands, for example in the Punjab, Sonora and Luzon, nitrogen applications are as high as in the developed countries. At these rates, the consequent nitrate levels in aquifers and drinking water can be in excess of the WHO upper limit of 50 mg/ litre of nitrate and likely to pose a health hazard. Contamination is not confined to the Green Revolution lands, however. A survey of water from 3,000 dug wells in Indian villages revealed that about 20 per cent contained nitrates in excess of the WHO limit [15]. But in most of these cases the nitrates were derived from human or livestock waste, rather than fertilizers.

In theory, high nitrate levels in drinking water could be responsible for increases in certain cancers [16]. Nitrates are converted to nitrites in the body and will combine with substances, common in the diet, known as amines to produce nitrosamines, which are known to be carcinogenic. The question is whether this actually happens and the answer appears to be no, at least in the developed countries. A study of the town of Worksop in England in 1973, which had high levels of nitrate, over 90 mg per litre, in its drinking water, revealed a higher incidence of gastric cancer than in neighbouring towns with low nitrate levels in their drinking water [17].

This finding attracted considerable attention and stimulated moves to restrict fertilizer use. But the study had left out a number of significant factors, including the high proportion of coal miners in the Worksop population. Miners and their wives are more likely to contract gastric cancer than the general population. Subsequent re-examination of the data, and further studies elsewhere, have shown no evidence of a link between nitrate levels and gastric cancer. Indeed, countrywide analyses in the United Kingdom have revealed a reverse correlation – the incidence of the cancer is higher in areas with low nitrate in the drinking water [18]. Nitrates may play a role but other factors appear to be far more important.

There are probably greater grounds for concern in the developing countries. Gastric cancer is particularly common in countries with high rates of use of nitrogen fertilizers. Chile is the only country with natural deposits of nitrates, in the form of saltpetre, and has a long tradition of heavy fertilizer use and one of the highest rates of gastric cancer in the world [19]. But there is no correlation with nitrate levels in drinking water. A possible explanation, here and in Colombia, where there is also a high incidence of gastric cancer, is the ingestion of nitrates contained in vegetables and legumes, such as beans [20]. Another puzzle is the link between nitrogen applications and bladder cancer in Egypt. There is high incidence of this cancer in the intensively farmed lands of the Nile delta and it appears to be associated, in a complex manner, with high-nitrogen fertilizer applications (260 kg N/ha), bacterial contamination of the water and a prevalence of the parasitic disease, schistosomiasis [21].

A more worrying hazard is methaemoglobinaemia or the 'blue baby' syndrome, which affects infants in the first few months of life [22]. Nitrite in the bloodstream – derived from nitrate in drinking water – combines with the blood's haemoglobin to produce methaemoglobin. This interferes with the normal process whereby haemoglobin combines with oxygen and transports it to different parts of the body. When more than 10 per cent of the haemoglobin has been converted to methaemoglobin, infants start to show the typical symptoms of oxygen starvation. They begin to take on a slate-blue colour, first around the lips, then appearing on the fingers and toes. At 70 per cent or more the infant usually dies.

Deaths have occurred in the USA and in Europe, but in most instances only where the drinking water was also seriously contaminated with bacteria. Some strains of acutely pathogenic bacteria, such as *Vibrio* and *Salmonella*, which are a common cause of diarrhoea in developing countries, can reduce nitrate to nitrite. And since infants with diets deficient in vitamin C are more prone to methaemoglobinaemia, it should be very common in the developing countries. However, after a thorough search of the literature, Jules Pretty of IIED and I could only find one reference. The blood methaemoglobin of 500 infants younger than a year in a rural part of Namibia had been measured and about 8 per cent were found to have more than 5 per cent methaemoglobin and one, close to death, had a level of 35 per cent. All were being given nitrate–laden water drawn from wells contaminated by livestock droppings. Those infants who regularly took in vitamin C in their diets were less affected (Figure 6.5). Given the heavy contamination by livestock and human wastes of drinking water in developing countries, the growing use of nitrogenous fertilizers and the low intake of vitamin C levels in the diet, I would expect a much higher reported incidence of the blue–baby syndrome. It may be that the disease is hidden by the common occurrence of dysentery and other disorders.

Fertilizers are not directly toxic to wildlife, but they can damage wild plants by causing excessive growth, and can severely disrupt natural eco-systems. Perhaps the most damaging consequence of fertilizer contami-nation is the nutrient enrichment − eutrophication − of rivers, lakes and coastal waters. Nitrates and phosphates which have run off the land are responsible for generating dense blooms of algae and of surface plants, such as water hyacinth and water cabbage. These shade out the underlying aquatic plants which, when they die and decompose, remove oxygen from the water, causing extensive fish kills. A thriving fish-pen industry in the

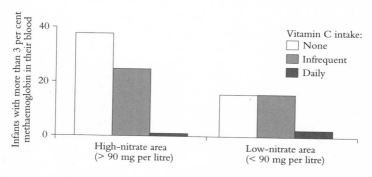

Figure 6.5 Blue-baby syndrome in infants in Namibia [23]

coastal lake of Laguna de Bay in Luzon is frequently damaged in this way and although the bulk of the nutrients entering the lake comes from human waste, a significant inflow is the fertilizer nitrogen from the surrounding intensively farmed ricefields [24].

The perceived health hazards from nitrates in drinking water have produced legally enforceable limits in many developed countries, and moves to restrict fertilizer application rates. Although the WHO limits are intended to be applied worldwide, they are unlikely to be implemented in the developing countries in the near future. Nevertheless, there may come a time when nitrogen fertilizer use is subject to the same kind of restrictions currently in force in the developed countries. In some respects, I believe this should be welcomed. While modern agrochemicals, such as fertilizers and pesticides, are undoubtedly instrumental in raising yields, there is a great deal of waste, in part caused by high levels of subsidies. In Indonesia subsidies for fertilizers had reached nearly 70 per cent, and for pesticides 40 per cent, of world prices in the early 1980s [25]. This accounts for the very rapid rise in fertilizer use in Indonesia, up by nearly 80 per cent between 1980 and 1985, and the much higher application rates, 75 kg/ha of arable land on average, compared with half and a third that amount in the Philippines and Thailand. In such situations restrictions on fertilizer use could result in more efficient application and less environmental damage, without loss of yield.

Increased, and inefficient, use of pesticides and nitrogen fertilizers produces severe pollution, but it is mostly local in its effect. Other agricultural pollutants have the potential for damage on a much larger scale. While industry is often to blame, agriculture is becoming a major contributor to regional and global pollution, producing significant levels of methane, carbon dioxide and nitrous oxide (Figure 6.6) [26]. These gases are generated by natural processes, but the intensification of agriculture in both the developed and developing countries has increased the rates of emission. Irrigated ricefields in Asia have grown 40 per cent in area since 1970, so contributing to increased production of methane and ammonia. Nitrous oxide emissions have grown in parallel with the use of nitrogen fertilizers. Ammonia and methane emissions have increased as a result of the intensification of livestock husbandry. And the clearance of forests and grasslands for arable land has raised the production of carbon and nitrogen oxides.

About half the world's emissions of methane are agricultural in origin. The gas is produced by specialized bacteria in environments that are free of oxygen (anaerobic), such as natural swamps and marshes and their agricultural equivalent, the wet ricefield or paddy [27]. Natural wetlands and ricefields each produce about 20 per cent of global methane emissions [28]. A second agricultural source is the anaerobic bacteria that live in the guts of ruminant animals. Cattle produce 35–55 kg/year, sheep and goats 5–15 kg [29]. Overall, ruminant animals are the cause of about 15 per cent of global methane emissions. A further 8 per cent results from the burning of agricultural wastes and fuel-wood, and from the burning of the savannas and fallow vegetation. Atmospheric methane has grown two and a half-fold since the beginning of the nineteenth century, as a consequence of increased agricultural activity and the burning and mining of fossil fuels.

Agriculture has a relatively smaller, but nevertheless important, impact on the global cycle of nitrous oxide (N_2O) [31]. It is produced naturally by the action of bacteria on nitrogen compounds in tropical soils,

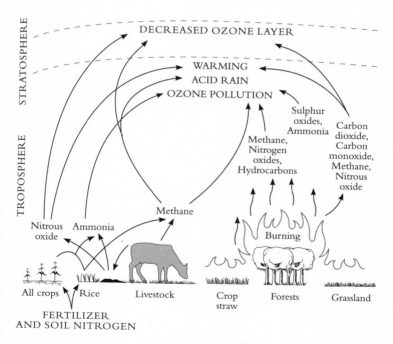

Figure 6.6 Global pollution caused by agriculture [30]

particularly those covered by undisturbed rainforest in the Amazon and elsewhere. This accounts for 20–30 per cent of total N_2O emissions. Bacteria will also liberate nitrous oxide from nitrogen fertilizers, some of the highest emissions coming from the breakdown of urea, the fertilizer most commonly applied in many developing countries. Fertilizers contribute 5–20 per cent of global emissions, which have increased by some 13 per cent since the time of the industrial revolution.

Most of the ammonia in the atmosphere comes from agriculture, principally from the volatilization of the urine and excreta of livestock – between 20 and 30 million tons a year worldwide [32]. The largest emitters are India and China, producing over 5 million and over 3 million tons respectively (the USA produces about 2·2 million tons) [33]. Ammonia emissions also result from fertilizer applications, and are particularly high from fertilized paddy fields [34].

One of the biggest sources of global air pollution is the burning of vegetation, in particular during the clearing of forest, scrub and grassland to make way for agriculture [35]. When vegetation is burned, the accumulated phosphorus and potassium may be returned to the soil, while the carbon, sulphur and nitrogen are lost as gases to the atmosphere. It is not a new phenomenon, but the scale of land clearance has greatly increased over the past three decades. Large areas are cleared annually for shifting cultivation. Forest or scrub is cleared, the felled vegetation burned and the land planted with crops for a number of years until the nutrients are depleted, the weeds build up and yields decline. Cropping may last as much as ten years before the cycle is repeated, but on poor soils or in areas of high population density farmers may abandon cropping after only two to three years. Estimates are difficult to make: probably between 200 and 300 million people are supported by shifting cultivation. They clear 20–60 million hectares per year, burning between 1 and 2 billion tons of dry matter [36]. This total also includes the burning, every three to five years, of the dead vegetation of the African and South American savanna lands to enhance the growth of grasses and so improve grazing. During the dry season the South American *cerrado*, which extends over 40,000 km², is usually covered with a dense layer of smoke. In addition, 8–15 million hectares of forests and savannas are transformed each year into permanent cultivation and into pasture for livestock raising, as well making way for settlements and highways. The increasing cattle population alone accounts for about half of global deforestation, and it is as high as 90 per cent in the Amazon. A

96

US satellite in 1987 photographed some 8,000 separate fires, each at least a square kilometre in size, in the Amazon basin. The layer of smoke, lying between 1,000 and 4,000 metres above the ground, extended over several hundred square kilometres [37].

Globally 1 to 2 billion tons of carbon, in the form of carbon dioxide, are emitted from biomass burning [38]. This is equivalent to between 50 per cent and 100 per cent of the carbon dioxide resulting from the burning of fossil fuels. Biomass burning also produces large quantities of methane and nitrous oxide, and some 30 million tons of particulate matter. The amount of carbon dioxide in the atmosphere has grown by over 25 per cent in the past century and is continuing to grow.

In summary, agriculture is a highly significant and growing contributor to the total production of globally important gas emissions (Table 6.1). Individually or in combination, these gases are contributing to:

— Acid deposition;
— The depletion of stratospheric ozone;
— The build-up of ozone in the lower atmosphere; and
— Global warming.

Table 6.1 Agricultural contributions to total global gas emissions (percentages)

Contributor	Methane	Nitrous oxide	Carbon dioxide	Ammonia
Paddy rice	21	–	–	Unknown
Livestock and wastes	15	Unknown	–	80–90
Fertilizers	–	5–20	–	< 5
Biomass burning	8	5–20	20–30	Unknown
Total	44	10–25	20–30	90

The effects on the natural environment and on human well-being are well known and do not need to be repeated here, but in each case there are significant adverse effects on agriculture.

Acidification of the air, soil and water is primarily a consequence of the release of sulphur dioxide and nitrogen oxides from the burning of fossil fuels, but ammonia also plays a role [39]. According to research in Europe, ammonia may have contributed to forest die-back and to longer-term changes in natural plant communities, especially on nutrient-poor soils.

Many regions in the developing countries are also experiencing increased acid deposition, largely as a result of the combustion of fossil fuels. In southern China the widespread burning of high-sulphur-content coal, particularly in domestic stoves and small-scale industry, is producing damaging acid rain. Around Chongqing large areas of paddy have suddenly turned yellow following acid rainfall, and wheat in the same area has suffered acute injury and reduced yields [40].

Acid rain also results from the nitrogen oxides produced during forest and savanna clearance. In Venezuela the burning of 15 million hectares of savanna each year produces some 27,000 tons of nitrogen oxides, which generate acid rain during the first rains after the dry season [41]. Although there is, as yet, little evidence of damage in tropical and subtropical countries, they may be more at risk from acidification, because their soils are often already acidic and unable to buffer any further additions to acidity.

The effects of the various gases produced by agriculture on ozone are seemingly paradoxical [42]:

− In the lower layers of the atmosphere − the troposphere − various gases, including methane and carbon monoxide, interact to increase ozone concentration; while
− In the upper atmosphere − the stratosphere, that extends from 10 to 50 kilometres above the earth's surface − some gases, notably methane and nitrous oxide, deplete ozone.

In both cases the chemical reactions involved are very complex. Ultraviolet radiation during daylight hours oxidizes carbon monoxide, methane and other hydrocarbons in the troposphere to produce ozone. At night the process is reversed, but the reversal is not necessarily complete and ozone accumulates. In the tropics and subtropics the high levels of ultraviolet radiation produce considerable accumulation, particularly close to cities as a result of the hydrocarbons emitted by motor vehicles. But ozone levels can also be high in rural areas. In the Amazon basin ozone is usually present at 10 to 15 ppb (parts per billion) but at up to 40 ppb during the clearing and burning season [43]. A similar effect occurs when savannas are burned.

High levels of ozone in the lower atmosphere will damage plants. Some crop plants are resistant; others, such as the legumes, are very sensitive. Experiments on rice have shown that ozone reduces the height of the plants and the weight of the seeds, and increases the number of sterile panicles, so reducing yields [44]. Around Tokyo, where ozone concentrations regularly reach 200 ppb, losses of rice could be as high as 30 per cent, and recent experiments have suggested high-yield losses in developing countries at much lower levels of ozone [45]. Nigel Bell of Imperial College and colleagues from the University of the Punjab in Pakistan used open-top chambers to grow rice and wheat under as natural conditions as possible, at a site to the south of Lahore. When the ozone polluting the air was filtered out, the yields were over 40 per cent greater for some varieties, a much higher effect than has been recorded in the developed countries. The reasons are not clear, but there may be a synergistic effect between ozone and other pollutants such as nitrous oxide.

About 90 per cent of all the ozone in the atmosphere is in the upper layers: the stratosphere. There, ozone is produced by the action of ultraviolet light on oxygen, and is broken down in a process that involves a variety of molecules, including chlorine, the hydroxyl ion and nitric oxide [46]. A major cause of the recent destruction of ozone is the chlorine produced by the chlorofluorocarbons used as aerosol propellants and refrigeration coolants, but agriculture is also a culprit since the hydroxyl ion is derived from methane, and nitric oxide from nitrous oxide.

The processes are extremely complex and still not fully understood. To cite just one complication: methane may also be important in removing the chlorine, by transforming it to hydrochloric acid which is then washed out of the stratosphere by rainfall. Thus methane plays several important, and contrary, roles in ozone chemistry, producing ozone in the troposphere, destroying ozone in the stratosphere, and, at the same time, also removing other destructive agents. Scientists have tried to model these processes, but the computations involved are daunting. A fully comprehensive model would contain over thirty constituents and 200 reactions and have to be replicated for different altitudes and latitudes and for different times of the year. Inevitably the models are simplified, leaving the results open to considerable debate. Until the 1980s, the models predicted little effect of increasing man-made emissions. But in 1985 the British Antarctic Survey, led by Joe Farman, reported the existence of a large springtime 'hole' in the ozone layer over the Antarctic [47]. Direct observations had picked up what the models had failed to predict. Since then the hole has remained constant in extent but has deepened each spring. It has also begun to appear in the northern hemisphere, extending over northern Europe and North America. So far, however, there have been no signs of loss over tropical regions.

The existence of the ozone layer is important for life on the planet since it serves to screen out the incoming ultraviolet (UV) radiation. High levels of UV light cause skin cancers in human beings and damage to plants [48]. Cereals are relatively tolerant, but many other crops, including legumes, squashes and cabbages, are easily damaged. Yields are reduced and, in soybean and potatoes, the oil and protein content is lowered.

More serious than the depletion of the ozone layer is the long-term effect of agricultural activities on the global climate. Together with water vapour, several of the gases emitted to the atmosphere by agriculture – carbon dioxide, methane and nitrous oxide – absorb the heat being radiated out from the earth's surface. The effect on the global climate is similar to the trapping of heat in a greenhouse. There is no doubt we are witnessing a significant increase in average global temperature, which began in the early part of this century and is greater than any rise in the last 500 years (Figure 6.7). This global warming may be a natural occurrence, but the best-informed opinion, in particular that of the Intergovernmental Panel on Climate Change (IPCC), believes it has been a direct consequence of the growing emissions of the so-called greenhouse gases [49].

Carbon dioxide is the most important greenhouse gas, contributing more

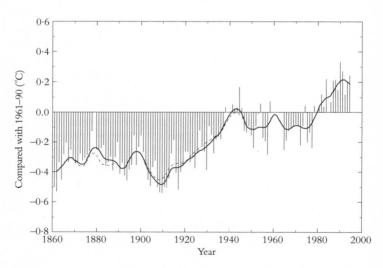

Figure 6.7 The rise in global temperature [50]

than half of global warming, with methane and nitrous oxide responsible for a fifth in the 1980s [51]. According to the IPCC, the increases in all these gases have already caused a temperature rise of 0·3−0·6 °C during the last one hundred years. There has been a sharp upward trend over the past ten years, although 1992 and 1993 were relatively cooler as a consequence of the material spewed out to the atmosphere from the volcanic eruption of Mt Pinatubo in 1991. The years subsequent to 1993 have been the hottest on record.

If nothing is done to control greenhouse-gas emissions, the global average temperature could rise by a further 0·4 °C by the year 2020 and as much as 2·0 °C by the end of the next century [52]. However, these are average global temperature figures. The temperature changes and their effects will vary from place to place and in ways that are not yet fully predictable. The greatest temperature changes will be at high latitudes, but water availability may worsen at lower latitudes. How far these changes will affect agriculture is difficult, at this stage, to assess. In the most thorough review to date, Martin Parry of University College, London, believes that heat and water stress may result in yield reductions, especially in the low latitudes, where most of the developing countries are situated [53]. By contrast, in the middle and high latitudes the increased CO_2 will have a physiological effect encouraging crop growth, particularly of so-called C_3 crops like wheat, barley, rice and potatoes [54]. On average, a doubling of CO_2 produces a 30 per cent increase in yield in these crops [55]. Combined with higher average temperatures this may increase production of grain and other crops in the developed countries.

Sea levels are also expected to rise, initially from the thermal expansion of the oceans and eventually, perhaps, as a result of the melting of the polar ice caps. The best estimate, on current rates of global warming, is that the average sea level will have risen by up to 15 centimetres by 2020 [56]. This is not a great amount but it could lead to a greater risk of flooding in countries such as Bangladesh, where much of the cultivation is precariously sited in the delta of the Brahmaputra.

The most serious consequences are some time in the future, towards the end of the next century. Many doubt whether a rise of a further 0·4 °C by the year 2020 is likely to have a major effect on agricultural production [57]. IFPRI has ignored any effects of global warming in its 2020 modelling exercise. But there may be grounds for concern. There has already been a significant fall in precipitation in the subtropics and tropics since 1960 (Figure 6.8). If this trend continues, it could soon have severe effects on agricultural production, particularly in the semi-arid tropics. The latest projections by the IPCC indicate a reduction in the rainfall produced by the South Asian monsoon as a result of global warming.

Another consequence of global warming may be a greater variability in the weather and a higher incidence of extreme weather conditions, with unpredictable effects [58]. Floods, droughts, hurricanes, extremely high temperatures and severe frosts may become more common. In the developing countries rainfall may become more variable, possibly with a greater frequency of heavy rainstorms creating flooding and exacerbating soil erosion [59]. The rainy season might also shorten, reducing the pre-monsoon rains that are crucial for crop germination. There are some recent signs of a greater frequency of such extreme events. In both North America and Sub-Saharan Africa there were severe droughts in the 1980s and they may be responsible for the increased variability in cereal production over the same period in these regions (Figure 6.9).

In the public mind, pollution is a matter of increasing concern, particularly in the developed countries, and while industry is seen as the main culprit, there is a growing realization of the significant role played by agriculture. Demands for 'organically produced' foods in the developed countries are beginning to have an impact on agricultural practices there, and some of the consequences may soon be felt in the developing world. Most serious are likely to be restrictions on inputs and practices which are, or at least seem to be, polluting. Developing countries reliant on exports

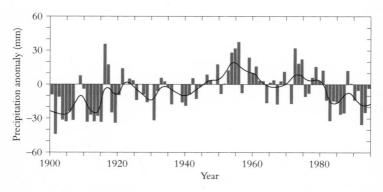

Figure 6.8 Changes in global precipitation [60]

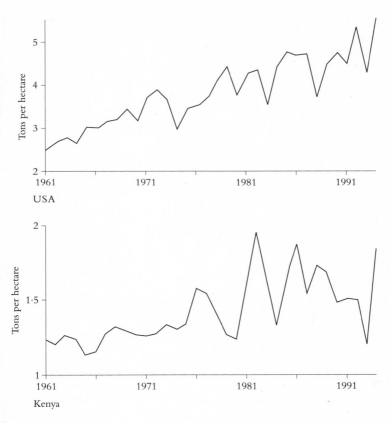

Figure 6.9 Increased variability of cereal yields in Kenya and the USA

of high-value fruits and vegetables will find themselves under pressure to reduce or eliminate pesticide use from these crops. There may be pressures for enforcement of the WHO global limits on nitrates in drinking water, leading to a curbing of fertilizer use. Multinational manufacturers of agrochemicals may also curtail research and investment in new products, even safer products, if they believe sales will be reduced because of public perception. Research on and development of insecticides have significantly declined in recent years, partly because of the prohibitive costs of developing new products that meet world standards [61]. One consequence is a paucity of new (and much needed) selective compounds, particularly for crops which have relatively small global production.

The threat of global pollution, still some way off in its direct effect on agricultural production, may also begin to produce restrictions on agricultural practices unless ways can be found of minimizing the emissions of the damaging gases.

Notes

[1] G. R. Conway, U. Lele, J. Peacock and M. Piñeiro, 1994, *Sustainable Agriculture for a Food Secure World*, Washington DC, Consultative Group on International Agricultural Research/Stockholm (Sweden), Swedish Agency for Research Cooperation with Developing Countries

[2] G. R. Conway and J. N. Pretty, 1991, *Unwelcome Harvest: Agriculture and Pollution*, London, Earthscan; D. Bull, 1982, *A Growing Problem: Pesticides and the Third World Poor*, Oxford, Oxfam

[3] Bull, op. cit.

[4] N. Alexandratos (ed.), 1995a, *World Agriculture: Towards 2010. An FAO Study*, Chichester (UK), Wiley & Sons

[5] S. Postel, 1987, *Defusing the Toxics Threat: Controlling Pesticides and Industrial Waste*, Washington DC, Worldwatch Institute (Worldwatch Paper No. 79)

[6] Alexandratos, op. cit.

[7] GAO, 1989, *Export of Unregistered Pesticides is Not Adequately Monitored by the EPA*, Washington DC, US General Accounting Office

[8] R. S. Levine, 1986, 'Assessment of mortality and morbidity due to unintentional pesticide poisonings', working paper presented to the Consultation on Planning Strategy for the Prevention of Pesticide Poisoning, Geneva (Switzerland), (Doc. No. WHO/VCB/86.926)

[9] M. E. Loevinsohn, 1987, 'Insecticide use and increased mortality in rural central Luzon, Philippines', *Lancet*, 13 June 1987, pp. 1359–62

[10] The mortality rate is age adjusted and graphed as a three-year moving average. – Loevinsohn, op. cit.

[11] A. Rola, 1989, *Pesticides, Health Risks and Farm Productivity: A Philippine Experience*, Los Banos (Philippines), University of the Philippines (Agricultural Policy Research Program Monograph, 89-01)

[12] All differences are significant except for the neurological impairments. A. Rola and P. Pingali, 1993, *Pesticides, Rice Productivity, and Farmers' Health: An Economic Assessment*, Los Banos (Philippines), International Rice Research Institute

[13] The Acceptable Daily Intake (ADI) is determined by the FAO/WHO Codex Alimentarius Commission and is calculated at 100 times lower than the amount

which would cause the least detectable effect if that level were ingested regularly throughout life.

[14] Conway and Pretty, op. cit.

[15] B. K. Handa, 1983, 'Effect of fertilizer use on groundwater quality in India', in International Association of Hydrological Sciences, *Groundwater in Resources Planning*, Vol. II, pp. 1105–9 (IAHS publ. No. 142)

[16] Conway and Pretty, op. cit.

[17] M. J. Hill, G. Hawksworth and G. Tattershall, 1983, 'Bacteria, nitrosamine and cancer of the stomach', *British Journal of Cancer*, 28, 562–7; J. M. Davies, 1980, 'Stomach cancer mortality in Worksop and other Nottinghamshire mining towns', *British Journal of Cancer*, 41, 438–45

[18] S. S. A. Beresford, 1985, 'Is nitrate in the drinking water associated with the risk of cancer in the urban UK?' *Journal of Epidemiology*, 14, 57–63

[19] R. Armijo and A. M. Coulson, 1975, 'Epidemiology of stomach cancer in Chile – the role of nitrogen fertilizers', *International Journal of Epidemiology*, 4, 301–9

[20] R. Armijo, A. González, M. Orellana, A. N. Coulson, J. W. Sayre and R. Detels, 1981, 'Epidemiology of gastric cancer in Chile; II – Nitrate exposures and stomach cancer frequency', *International Journal of Epidemiology*, 10, 57–62; E. Fontham, D. Zavala, P. Correa, E. Rodríguez, F. Hunter, W. Haenszel and S. R. Tannenbaum, 1986, 'Diet and chronic atrophic gastritis; a case-control study', *Journal of the National Cancer Institute*, 76, 621–7

[21] R. M. Hicks, M. M. Ismail, C. L. Walters, P. T. Beecham, M. F. Rabie and M. A. El Alamy, 1982, 'Association of bacteriuria and urinary nitrosamine formation with *Schistosoma haematobium* infection in the Qalyub area of Egypt', *Transactions of the Royal Society of Tropical Medicine and Hygiene*, 76, 519–27; Conway and Pretty, op. cit.

[22] Conway and Pretty, op. cit.

[23] M. Super, H. de V. Hesse, D. MacKenzie, W. S. Dempster, J. Du Plessis and J. J. Ferreira, 1981, 'An epidemiological study of well water nitrates in a group of SW African/Namibian infants', *Water Research*, 15, 1265–70

[24] R. B. Edra, 1983, 'Laguna de Bay – an example of a fresh and brackish water fishery under stress of the multiple-use of a river basin', *FAO Fish Report*, 288, 119–24

[25] World Bank, 1987, *Indonesia – Agricultural Policy: Issues and Options*, Washington DC, World Bank

[26] Conway and Pretty, op. cit.

[27] A. Holzapfel-Pshorn and W. Seiler, 1986, 'Methane emissions during a cultivation period from an Italian rice paddy', *Journal of Geophysical Research*, 91, 11803–14

[28] A. F. Bouwman, 1990, 'Land use related sources of greenhouse gases', *Land*

Use Policy, April 1990, pp. 154–64; R. Watson *et al.*, 1990, 'Greenhouse gases and aerosols', in Intergovernmental Panel on Climate Change (IPCC), *The Scientific Assessment of Climate Change*, Working Group 1, Section 1, Peer reviewed Assessment for Working Group 1, Plenary, 25 April 1990, Geneva (Switzerland), World Meteorological Office

[29] P. J. Crutzen, I. Aselmann and W. Seiler, 1988, 'Methane production by domestic animals, wild animals and other herbivorous fauna, and humans', *Tellus*, 38B, 271–84

[30] Conway and Pretty, op. cit.

[31] Watson *et al.*, op. cit.

[32] I. E. Galbally, 1985, 'The emission of nitrogen to the remote atmosphere', in J. N. Galloway, R. J. Charlson, M. O. Andreae and H. Rodhe (eds.), *The Biogeochemical Cycling of Sulfur and Nitrogen in the Remote Atmosphere*, Hingham, Mass., D. Reidel

[33] Conway and Pretty, op. cit.

[34] D. S. Mikkelson and S. K. De Datta, 1979, 'Ammonia volatilisation from wetland rice soils', in IRRI, *Nitrogen and Rice*, Los Banos (Philippines), International Rice Research Institute

[35] Conway and Pretty, op. cit.

[36] W. Seiler and P. J. Crutzen, 1980, 'Estimates of gross and net fluxes of carbon between the biosphere and the atmosphere from biomass burning', *Climatic Change*, 2, 207–47

[37] J. Rocha, 1988, 'Ozone fears as Amazon forest burns', quoting Paul Crutzen and Richard Stolarski, *Guardian*, 18 April 1988, p. 6

[38] Watson *et al.*, op. cit.; Bouwman, op. cit.

[39] Conway and Pretty, op. cit.

[40] D. Zhao and J. Xiong, 1988, 'Acidification in southwestern China', in H. Rodhe and R. Herrera (eds.), *Acidification in Tropical Countries*, Chichester (UK), (SCOPE Report No. 36), Wiley & Sons, pp. 317–46

[41] E. Sandueza, G. Cuenca, M. J. Gomez, R. Herrera, C. Ishizahi, J. Marti and J. Paolini, 1988, 'Characterisation of the Venezuelan environment and its potential for acidification', in Rodhe and Herrera, op. cit., pp. 197–256

[42] Conway and Pretty, op. cit.

[43] V. W. J. H. Kirchhoff, 1988, 'Surface ozone measurements in Amazonia', *Journal of Geophysical Research*, 93, 1469–76

[44] G. Kats, P. J. Dawson, A. Bytnerowicz, U. W. Wolf, C. R. Thompson and D. M. Olszyk, 1985, 'Effects of ozone and sulphur dioxide on growth and yield of rice', *Agriculture, Ecosystems, Environment*, 14, 103–17

[45] A. Wahid, R. Maggs, S. R. A. Shamsi, J. N. B. Bell and M. R. Ashmore, 1995, 'Air pollution and its impact on wheat yield in the Pakistan Punjab', *Environmental Pollution*, 88, 147–54; 1995, 'Effects of air pollution on rice yield in the Pakistan Punjab', *Environmental Pollution*, 90, 323–9

[46] UK SORG, 1988, *Stratospheric Ozone*, Stratospheric Ozone Review Group, Department of the Environment and the Meteorological Office, London, Her Majesty's Stationery Office

[47] J. C. Farman, B. G. Gardiner and J. D. Shanklin, 1985, 'Large losses of total ozone in Antarctica reveal seasonal ClO_X/NO_X interaction', *Nature*, 315, 207–10

[48] R. C. Worrest and L. D. Grant, 1989, 'Effects of ultraviolet-B radiation on terrestrial plants and marine organisms', in R. Russell-Jones and T. Wigley (eds.), *Ozone Depletion: Health and Environmental Consequences*, Chichester (UK), Wiley & Sons

[49] J. T. Houghton, L. G. Meira Filho, B. A. Callander, N. Harris, A. Kattenberg and K. Maskell, 1996, *Climate Change 1995: The Science of Climate Change*, Contribution of Working Group I to the second Assessment Report of the Intergovernmental Panel on Climate Change, Cambridge, Cambridge University Press

[50] Top graph: summer temperatures in the northern hemisphere derived from tree rings, ice cores and documentary records. Bottom graph: combined land–surface air and sea temperatures. – Houghton *et al.*, op. cit.

[51] J. Leggett, 1990, 'The nature of the greenhouse threat', in J. Leggett (ed.), *Global Warming: The Greenpeace Report*, Oxford, Oxford University Press

[52] Houghton *et al.*, op. cit.

[53] M. Parry, 1990, *Climate Change and World Agriculture*, London, Earthscan; C. Rosenzweig and M. L. Parry, 1994, 'Potential impact of climate change on world food supply', *Nature*, 367, 133–8

[54] C_3 and C_4 refer to different photosynthetic mechanisms. Most crops, especially in cooler and wetter habitats, are C_3. Plants such as tropical grasses, maize, sugar cane and sorghum are C_4.

[55] J. Reilly, 1996, 'Agriculture in a changing climate: impacts and adaptation', in R. T. Watson, M. C. Zinyowera and R. H. Moss, *Climate Change 1995: Impacts, Adaptations and Mitigation of Climate Change: Scientific-technical Analyses*, Contribution of Working Group II to the second Assessment Report of the Intergovernmental Panel on Climate Change, Cambridge, Cambridge University Press, pp. 427–67

[56] Houghton *et al.*, op. cit.

[57] T. Dyson, 1996, *Population and Food: Global Trends and Future Prospects*, London, Routledge

[58] R. W. Katz and R. G. Brown, 1992, 'Extreme events in a changing climate: variability is more important than averages', *Climate Change*, 21, 289–302; Reilly, op. cit.

[59] Parry, op. cit.

[60] Houghton *et al.*, op. cit.

[61] Conway and Pretty, op. cit.

7   Trends and Priorities

As the world progresses through the nineties, each year brings additional evidence that we are entering a new era, one quite different from the last four decades. An age of relative food abundance is being replaced by one of scarcity.

– Lester Brown, *Who Will Feed China?* [1]

The relatively favourable outcome of the models which I described in Chapter 2 is dependent on the growth in crop yields continuing at the rate experienced in recent years. If this turns out to be incorrect, market demand in the year 2020 will be satisfied at a much lower level of production and the hidden need of the hungry will rise, increasing the numbers of chronically undernourished. In the first part of this chapter I ask: Is the assumption of continuing yield growth sound? Is there evidence of environmental and other constraints placing a limit on future production?

At the outset, the importance of yield growth needs to be underlined. While there is considerable potential for expanding cropland in the developed countries, by bringing set-aside and conservation land back into production (see page 117 below), there is less room for expansion in the developing countries, at least in those countries likely to be most hard-pressed. The developing world contains about 2·6 billion hectares of land with potential for arable cultivation, of which about three-quarters of a billion hectares are currently cropped [2]. If the remaining 1·8 billion hectares were to realize the average present grain yield for the developing countries, i.e. 2·56 tons/ha, total production could increase by over 4·5 billion tons. In theory, this could ensure the world was well fed for many generations to come; but, for a variety of reasons, it is an impractical option.

The distribution of the potential new arable land is highly uneven. Over 90 per cent is in Sub-Saharan Africa and Latin America, and over one-third in only two countries: 27 per cent in Brazil and 9 per cent in Zaire (Figure 7.1). There is little room for expansion in South Asia and West Asia/North Africa. Nearly half of the 'potential' arable land in South Asia is occupied by cities and other human settlements.

Most of potential cropland lies distant from the main centres of population pressure. In Sub-Saharan Africa the highest densities – 100 or sometimes over 250 people per km² – are in Nigeria, and in the Ethiopian and East African highlands [3]. The density in Zaire is only 18 per km², but there is little likelihood of large-scale immigration from other parts of Africa. More feasible is resettlement within countries, people moving or being moved from densely populated regions to open up new lands – in the interior of Brazil or on the outer islands of Indonesia. Indeed, this is already under way. However, there are formidable problems. The quality of the land is poor, three-quarters of the potential in Sub-Saharan Africa and Latin America being subject to soil and terrain constraints (Figure 7.2). These can be overcome in many situations, but only with very careful attention to the maintenance of soil fertility and structure and the correction of various deficiencies and toxicities [4].

Much of the 'unexploited' potential arable land is under primary forest, in the Amazon and Congo basins and on the island of Borneo. Primary forest gives the impression of immense, luxuriant growth, but the productivity can be fragile, depending on the recycling of nutrients from the forest canopy

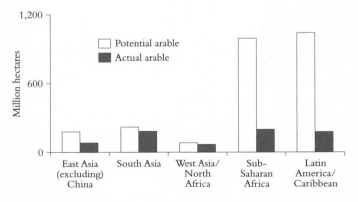

Figure 7.1 Potential arable land in the developing countries [5]

109

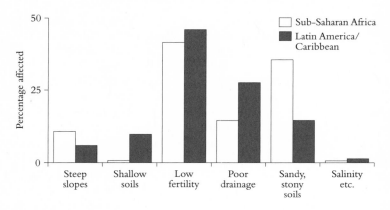

Figure 7.2 Percentage of potential cropland suffering from soil and terrain constraints [6]

to the litter on the forest floor and back again directly through the tree roots. Often there is no deep soil fertility and micronutrient deficiencies are common. In the 1970s I worked with Indonesians on a transmigration scheme in the Upang delta of Sumatra. Having cleared the swamp forest, they were growing rice in highly brackish water, attempting to control the salinity using a complicated system of flushing out the salts. But subsequently there has been rapid deterioration of the peat soils, and the original objective of double rice-cropping has had limited success. Since 1969 over 600,000 families have been resettled from Java to Sumatra, Kalimantan, Sulawesi and Irian Jaya, and the plan is to resettle some 10 million by the turn of the century [7]. Most of the land, though, is unsuitable for rice and other annual crops. It is acidic, with low organic and mineral content, with poor drainage and liable to heavy erosion. Forest land, if it is to be cultivated, is better replaced usually with tree crops, such as oil palm and cacao, than with cereals. In another transmigration project in Kalimantan, on the island of Borneo, where the settlers had been directed to grow rice, they were inserting rows of coconut palms in the ricefields, gradually turning their farms into small coconut plantations, although this was against government policy.

Nikos Alexandratos of FAO has carried out a detailed analysis of the potential for expansion in crop area in the developing countries, taking into account these and other factors, and estimates that by 2010 there will be an increase of only about 90 million hectares, about 5 per cent of the 'potential'. Nearly three-quarters of the extra cropped land will be in Sub-Saharan Africa and Latin America, with some expansion in East Asia (mostly in Indonesia and Cambodia), but virtually none elsewhere. In Asia, as a whole, the currently available 0·15 hectares of cropland per person will fall to a mere 0·09 hectares by 2020. And, according to estimates by Lester Brown of the Worldwatch Institute, China's land under cereals will fall to below 0·05 ha per person.

Where irrigation is available, the harvested area can be significantly increased by growing more than one crop a year. On average, one and a half crops are now grown on every arable hectare in China, which is partly why rice yields are higher than in India. Alexandratos estimates that, by 2010, the harvested area could increase by a further 34 million hectares over and above the 90 million hectares of new land, as a result of increased multiple-cropping in Asia, and a shortening of the fallow in Sub-Saharan Africa. An increase in the developing-country harvested area of 124 million hectares would make a significant contribution to total agricultural production, but it would not be easy. Currently uncultivated land tends to have severe constraints and increased multiple-cropping requires standards of irrigation availability that are rarely to be found. In most countries of the developing world, greater production will have to come from yield growth. Since the 1950s some 90 per cent of the increase in cereal production has been due to greater yields [8]; in future the proportion will need to be even higher.

There are, at least in theory, no major physiological, genetic or agronomic constraints to achieving annual yield gains of 2 per cent or more. Conventional plant-breeding techniques, increasingly augmented by genetic engineering (see the next chapter), should be able to produce improved plant types capable of significantly higher yields in all parts of the world. According to calculations by a group of Dutch scientists, the most productive lands in the world are potentially able to deliver annual yields of over 25 tons of grain equivalents per hectare, given optimum sunlight, water and nutrients, and freedom from pests and diseases [9]. Indeed, there are a few records approaching this level. A single crop of maize in the USA has yielded nearly 24 tons; in China, wheat followed by two crops of rice has produced over 24 tons [10]. Only some land is of this high

111

quality and is situated where sunshine and water are plentiful, but much of it is in the developing countries. The Dutch group have assessed the lands of each of the world's continents in terms of potential productivity, taking account of the solar radiation, the length of growing season and other factors, and have produced for each an average theoretical maximum yield (Figure 7.3). Perhaps rather surprisingly, the highest maxima occur in the continents containing the developing countries; but, not surprisingly, it is in these regions where the gaps between theoretical and realized yields are greatest.

One way of achieving these high yields is to apply more fertilizer. Present levels of application vary between the developing regions, but are well below those common in the developed countries – 402 kg/ha in Japan and 614 kg in the Netherlands (Figure 7.4). And while fertilizer consumption grew rapidly in the 1970s and early 1980s, it has declined in the developed countries over the past decade and is now showing signs of deceleration in the developing countries (Figure 7.5).

Equally crucial to higher yields is the supply of water. Irrigated land accounts for some 16 per cent of the world's crop land, but produces 40

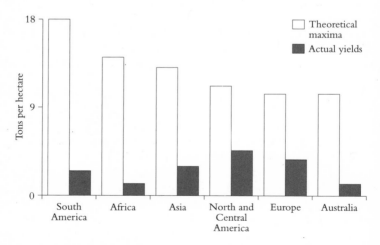

Figure 7.3 Theoretical maximum yields in grain equivalents and existing average grain yields for the continents of the world [11]

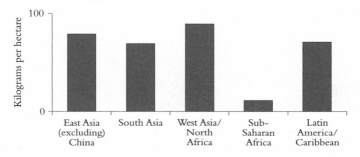

Figure 7.4 Average fertilizer application rates in the developing world [12]

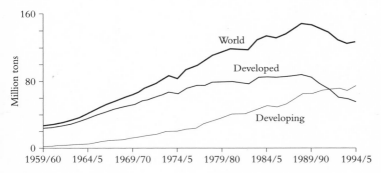

Figure 7.5 Total fertilizer use in the developed and developing countries [13]

per cent of the world's food [14]. In Chapter 4 I described the early recognition of the importance of assured, well-regulated water supply if the new varieties were to achieve their potential. Nearly 1,000 large dams came into operation, worldwide, every year in the 1950s and 1960s, but these high rates of expansion have not lasted. In the 1990s the number of new large dams has fallen to 260 a year [15]. Although in India the growth of irrigated land continues with only a slight deceleration, there has been a significant slowdown in the Philippines and Pakistan, while in China, Mexico, Malaysia and Egypt a plateau was reached in the mid-1970s

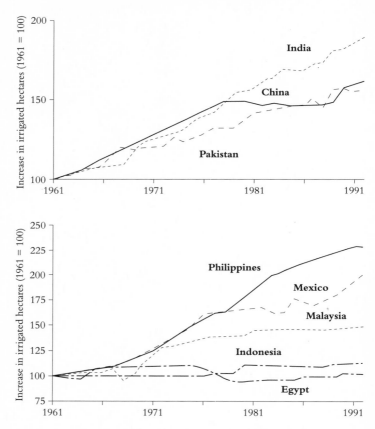

Figure 7.6 The growth in irrigated land in selected developing countries

(Figure 7.6). Only recently are there signs of a growth again in China and Mexico.

Even at its peak, the growth was less than the rate of population increase, and the decline in the amount of irrigated land per person in the developing countries has accelerated since the late 1970s (Figure 7.7).

According to the World Bank, the current 170 million hectares of irrigated land in the developing countries could be expanded by nearly 60 per cent, most of the potential lying in Asia [16]. In the 1980s China was

114

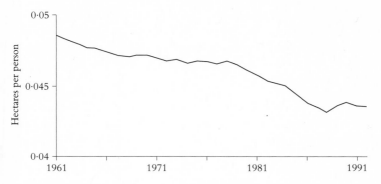

Figure 7.7 Irrigated land per person in the developing countries

building 183 dams over 30 metres in height and India a further 160 dams.
A number of massive projects, including the Narmada Valley in India and
the Three Gorges on the Yangtse river in China, are currently being
undertaken. But the investment is considerable and, as the experience of
recent irrigation projects makes clear, they face formidable technological,
environmental and social constraints (see Chapter 13). Not the least of the
difficulties is increasing competition for water, to provide irrigation not only
for agriculture but also to satisfy growing domestic and industrial demands.
Agriculture accounts for some 70 per cent of managed water resources, 21
per cent going to industrial and 6 per cent to domestic use, and agriculture's
share is likely to fall [17]. Cities such as Jakarta and Cairo are situated on good
agricultural land and take both land and water away from crop production.
Agriculture loses out because industry and domestic users can afford to pay
more for their water and can wield greater political influence.

Alexandratos estimates that by 2010 only some 23 million hectares may
be added to the developing countries' stock of irrigated land. And this
assumes a restitution or replacement of land lost to salinization and other
forms of degradation. At present some 25 million hectares globally suffer
from salinization and 2 million hectares of saline land are being added each
year (see Chapter 13) [18]. According to David Seckler of the International
Irrigation Management Institute in Sri Lanka, 'the net growth of irrigated
area in the world has probably become negative' [19]. The most significant
gains are likely to come from the small-scale harnessing of water, especially
in Sub-Saharan Africa (Chapter 13).

Other constraints to higher yields are micronutrient deficiencies and other soil toxicities, aridity and waterlogging, and pest, pathogen and weed attack (see Chapters 11–14). Solutions are usually technically feasible but at a cost in terms of capital or research investment, or in forgone opportunities, and with the risk of the solutions creating further problems. The field replication of yield gains achievable in the laboratory or in an experimental plot is limited less by narrowly biological constraints and more by broader environmental, economic and social factors. The question is whether these broader constraints are now beginning to have a serious effect.'Are there signs of a plateau in production or yields?

At a global level, growth in world cereal production has slowed from a rate of about 3·5 per cent in the 1960s to less than 3·0 per cent in the 1970s and to below 2 per cent in the 1980s [20]. The decline has been most marked since the mid-1980s, down to about 1 per cent. As a consequence, per capita cereal production has stopped growing and, indeed, has begun to decline (Figure 7.8). On the face of it, this is a very worrying trend. It has excited much comment – an example is the quote from Lester Brown of the Worldwatch Institute at the beginning of this chapter [21]. However, the situation is complicated and the figures need careful analysis.

Most of the reduction in the rate of growth has occurred in the developed countries, which contribute one-half of the world's cereal harvest [22].

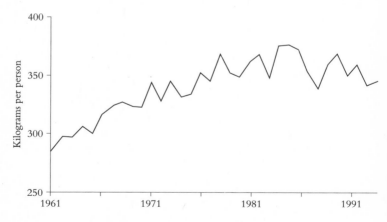

Figure 7.8 World per capita cereal production 1961–94

Faced with declining prices, stagnant exports and accumulating stocks, they have cut back on production and are largely responsible for the fall of 38 million hectares in the world's harvested area of cereals since the peak of 1981 (Figure 7.9) [23]. The cuts have taken various forms. In the USA, the passing of the Food Security Act in 1985 provided subsidies for cereal exports and initiated a process of reducing support costs, especially for maize and wheat. The arable land so released has been turned over to conservation schemes, such as the Conservation Research Program, under which some 20 million hectares were set aside in 1987 and 1988 [24]. In Europe there have been similar programmes of 'set-aside' and the substitution for cereals of alternative crops, such as rape and flax. Argentina, another major exporter, faced with subsidized competition from the USA and the European Union, has reduced its harvested area from some 11 million hectares to just over 8 million, much of the land being switched to soybeans. The largest reduction has been in the former Soviet Union, despite it being a major importer, in part through withdrawal of cultivation on marginal lands, where erosion has become a serious problem (Figure 7.10).

However, the decline in developed–country production is only part of the explanation for the lower per capita levels overall. As I indicated in Chapter 4, the impact of the Green Revolution has been very uneven

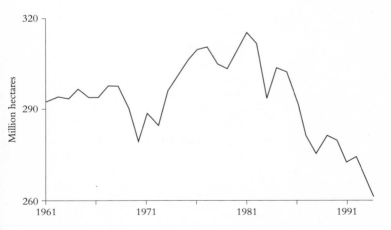

Figure 7.9 Fall in harvested area of cereals in the developed world

117

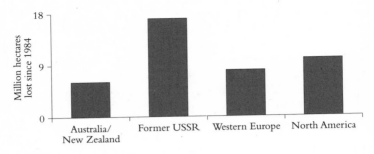

Figure 7.10 Fall in area of harvested cereals in developed countries in 1980s

geographically (Figure 7.11). Asia has been the greatest beneficiary, by far. This has been partly due to the greater availability of irrigation, partly because the two principal Green Revolution cereals – wheat and rice – are the main staples for most of the people of Asia, and partly because – in Indonesia, Malaysia, Thailand and the Philippines, as well as India and Pakistan – there exist governments willing and able to make and direct the necessary investments, including the necessary research structure which, through further breeding and selection, can adapt the new varieties to local conditions.

There has been less progress in raising production in Latin America. Part of the reason has been the marked dual nature of the agricultural economy. Since the 1960s the traditional haciendas have been transformed into large, capital–intensive farms, concentrating on the production of export crops. Food production has been left to small farmers, who have been slow to adopt the new technologies. Irrigated land in the 1950s totalled a mere 6·5 million hectares and has only increased by a further 10 million hectares. And two of the primary food staples of the region – potatoes and cassava – were not those first targeted by the Green Revolution plant breeders. Over Latin America as a whole, grain production has increased, but primarily as a result of the colonization of new land. Per capita production rose throughout the 1960s and 1970s, but has fallen sharply since 1985, owing to the contraction in Argentine wheat production. Remove the figures for Argentina and the fall in Latin America's per capita production remains but is much less great [25].

118

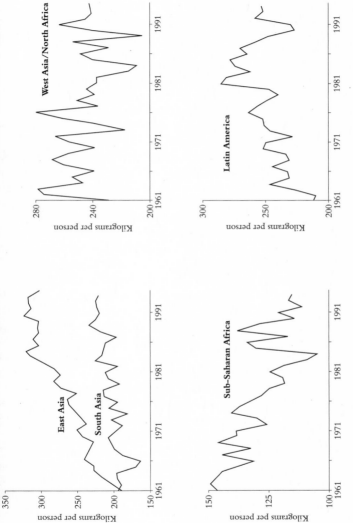

Figure 7.11 Cereal production per capita in the developing world

In West Asia/North Africa per capita production fluctuated about a downward trend line in the 1960s and 1970s. Since 1985, however, there has been a significant increase in production largely as a result of the expansion of cropped area, involving reclamation and irrigation of steppe and desert, including a major venture in wheat cultivation in Saudi Arabia based on the mining of deep fossil water.

The Green Revolution has been least successful in Sub-Saharan Africa. Cereal yields have changed little over the past forty years and cereal production per capita has steadily declined. As in Latin America, most cereals are produced by smallholders, and less than a quarter of food output is sold off the farm [26]. The amount of land that is irrigated is tiny – only 3 million hectares in 1960 and still only 6 million now, less than 5 per cent of the arable land. Half the increase in cereal production has come from expansion of land, either by opening up marginal lands or by reducing the fallow period. The important food staples over much of the continent are maize, millets, sorghums and root crops, for which, with the exception of maize, few high-yielding varieties have been produced.

In summary, while production per capita continues to grow in South Asia and, in recent years, in West Asia/North Africa, there is a slowing of growth in East Asia and Latin America and a continuing rapid decline in Sub-Saharan Africa. These figures, however, refer to cereal production; a more important question is what is happening to yields. To answer this we have to look at country data and, in particular, at the recent yield trends for the Green Revolution lands.

In South Asia, the growth of Pakistan's wheat yields slowed in the 1980s but in India there has been no slackening (Figure 7.12). Rice yields continued to grow in the 1980s in India and Bangladesh, and at a somewhat higher rate in India, but there are signs of a plateau in both countries over the last five years. In East Asia, the slowing down of yield growth is more apparent. In both the Philippines and Indonesia yields are growing at a slower rate than they did in the 1970s, and this is also true of rice and wheat in China.

Similar slowdowns are apparent in wheat yields in Latin America and North Africa. The growth of maize yields has remained low, although in Mexico and other countries of Latin America there has been a sudden increase in yields, after a plateau in the 1980s (Figure 7.13). Most of the yield gains in maize in Sub-Saharan Africa were in the 1950s and early 60s. Subsequent yield increases have been very erratic. Yields of millet

120

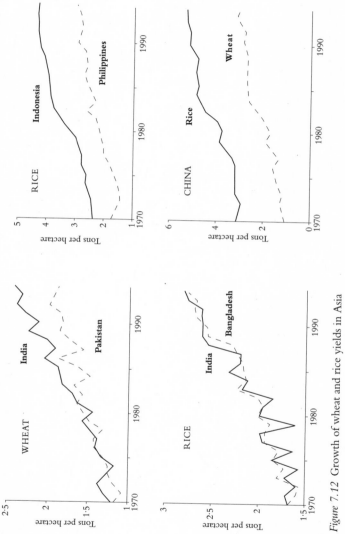

Figure 7.12 Growth of wheat and rice yields in Asia

and sorghum have remained more or less constant, partly because of their replacement on the better lands by maize.

There are several possible reasons for the widespread slowing of yield growth, and it is not clear which is the most significant. In the early years of the Green Revolution, plant breeding delivered steep increases in the maximal potential yields of each new variety to come on stream. Subsequently, the increases have been much less as breeders have concen-

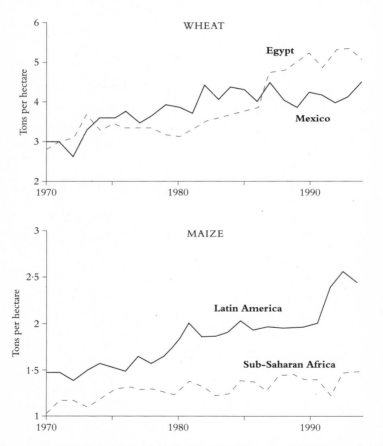

Figure 7.13 Growth in wheat yields in Egypt and Mexico and maize yields in Latin America and Sub-Saharan Africa

trated on other characteristics, such as pest and disease resistance (see the next chapter). This is unlikely to be a cause of slowing average yield growth, however, since average yields are well below the experiment-station potentials. A more significant, and more commonly accepted, cause is the spread of the new varieties on to more marginal lands, to which they are less suited and where yield gains are considerably more difficult.

More worrying in the long term is the evidence from a number of experimental sites of declining yields under intensive cropping on some of the better lands. Part of the reason seems to be the unexpected importance of micro-element deficiencies and toxicities. In the Chiang Mai valley in northern Thailand (Chapter 10) rice yields under double- and triple-cropping, in rotation with vegetables and other crops, grew rapidly in the early 1970s but then declined to a third of their peak level [27]. Farmers in the valley claim their land has become 'lifeless'. The soils have undergone considerable physical and chemical changes under intensive cropping; in particular there has been a growing problem of boron deficiency. This does not seem to be an isolated phenomenon [28]. At IRRI's research station in the Philippines there has been a considerable fall in yields in a long-term rice experiment that has run since the 1960s. The decline has been greatest in the yield of the original 'miracle rice' variety IR8, which has suffered from an increasing build-up of rice pests, but it is also apparent in the performance of the more resistant, highest-yielding varieties in the annual trials (Figure 7.14). The cause seems to have been the build-up of zinc deficiency and boron toxicity, a consequence of drawing the irrigation water from alkaline, high-boron wells. There have also been rice yield declines under intensive cultivation at many other sites in the Philippines, mostly due to growing phosphorus and potassium deficiency [29]. In the Indian Punjab there are increasing signs of zinc deficiency, as well as of deficiencies of sulphur, manganese and copper. In most cases, it is possible to correct for such deficiencies.

More serious are other adverse trends, where reversal is either impossible or likely to be very costly. In the Punjab, yields have grown dramatically since the 1950s. Ninety per cent of the arable land is irrigated, nitrogen fertilizers are applied on average at over 100 kg/ha and the cropping intensity is nearly 180 per cent. In the late 1980s rice–wheat rotations in some parts of the Punjab were delivering yields of 13·4–14·5 tons/ha – not far from the maximum biological potential (Figure 7.3). Wheat yields are still growing, but this achievement is now being seriously

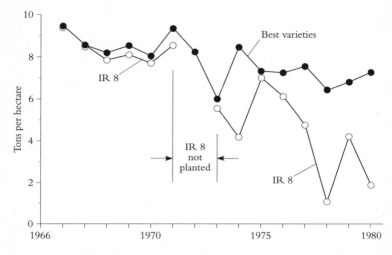

Figure 7.14 Yield declines at IRRI in the dry season (fertilizer applied at 120 kg/ha) [30]

threatened [31]. Of greatest concern is the growing scarcity of water. According to several estimates, good-quality water availability in the state is about 25 million acre feet, but the demand of the existing cropping intensity is about 37 million acre feet. There are some three-quarters of a million tube-wells drawing water at greater than the recharge rate and the Directorate of Water Resources has declared 82 blocks as hydrologically critical. In the most intensively cultivated districts of Patiala and Ludhiana, the groundwater table has fallen to a depth of 9–15 metres and is falling at about a half a metre a year. These two districts and the similarly affected district of Sangrur contribute nearly 40 per cent of the state's wheat and rice. As the water falls below 15 metres the tube-wells are being replaced by submersible pumps, at considerable expense. Salinization is also serious, affecting 9 per cent of the total cropped area, and nearly 300,000 hectares in south-west Punjab; some 100,000 hectares are also waterlogged. Reclamation is possible, but again at a high cost. This and other, albeit largely anecdotal, evidence from Luzon, Java and Sonora suggest there are serious and growing threats to the sustainability of the yields and production of the Green Revolution lands.

Fortunately there are few signs of declining yields in the developed countries (Figure 7.15). Cereal yields in North America and Western Europe have grown from about 4 tons/ha to 5 tons/ha over the past decade. There is no obvious technical reason why yields should not reach over 6 tons/ha by 2020. On the total grainland of some 118 million hectares currently under cultivation in North America and Western Europe, an extra ton would be sufficient to provide a further 118 million tons for export. However, there are environmental factors to consider. The extra production would require an increase in fertilizer application which, at least in Western Europe, would undoubtedly hit drinking-water restrictions. And there is the possible effect of global pollution, and global warming in particular. Atmospheric ozone levels are rising in Europe and may begin to seriously limit yield growth. Higher mean temperatures and elevated levels of CO_2 in North America and Europe will tend to raise yields, but if global warming is accompanied by greater climatic extremes, yields will become more variable, threatening global stocks and cereal trade stability. There have been several major droughts in North America during the 1980s, which could be the first signs of the effects of global warming. Yields in North America are still rising, but if the fluctuations worsen, the export potential in any one year would become more unreliable.

If global market demand were to rise above that predicted by the IFPRI model, one answer would be for land currently set aside in North America and Europe to be planted to cereals. Together these regions had an extra 25 million hectares cultivated with cereals in 1981. If brought back into grain production, at an average yield of 5 tons/ha, they could contribute a further 125 million tons for export. But again, there would be severe environmental constraints, not least in the increased levels of fertilizer use. Lester Brown claims that all of the 14 million hectares in the US Conservation Reserve Program is highly erodible, although Robert Paarlberg of Wellesley College, quoting various sources, challenges this assertion [32]. Developments over the past two decades in conservation tillage would make it possible for one-third of the reserve lands to be brought safely back into production, although average yields on these lands would probably be of the order of 3·5 tons/ha.

An unknown factor in future global grain production is the potential for yield growth in Eastern Europe and the former Soviet Union [33]. Currently yields are low and show little signs of rising, although in 1992

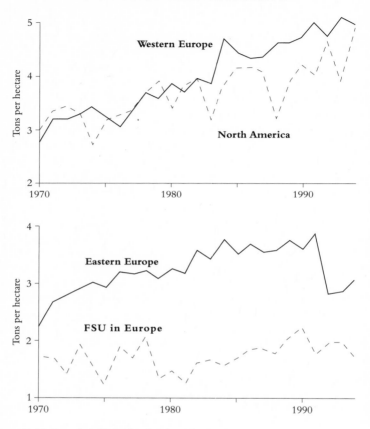

Figure 7.15 Cereal yields in the developed countries

yields in Hungary and the former Czechoslovakia were about 4 tons/ha (compared with only 2·5 tons in Poland). There is a potential for much higher yields throughout Eastern Europe, as there is in the former Soviet Union (where average yields in Russia are only some 1·5 tons/ha, but in the Ukraine are over 2·7 tons). At present there is great inefficiency in production and considerable post-harvest waste. Much-needed agricultural reforms will eventually make a difference, but are not likely to have a significant effect on global trade before the year 2020. Russia could bring back into production the over 44 million hectares of extra land under grain in the 1970s, but reportedly part of the reason for its withdrawal was the high levels of erosion. A more probable scenario is the recurrence of periods when these regions become significant net importers.

Another unknown in global food scenarios is the future trading require-ment of China. In 1995 Lester Brown published a highly provocative book in which he argued that soon China will have begun to take on some of the characteristics of the industrialized countries of East Asia, notably Japan; South Korea and Taiwan [34]. Their history shows a significant decline in grain production in the later period of industrializ-ation, not because of falling yields, but as a result of declining area planted to cereals as land is taken for other uses, notably transportation and urban development. Japan has lost over 50 per cent and South Korea and Taiwan over 40 per cent of their grain-harvested area over the last few decades. At the same time, as per capita incomes have grown, the populations have moved up the food chain in dietary terms, consuming more livestock products, especially poultry and pigs. This has resulted in an increased demand for grain for feedstuffs, which has had to be met by imports. By 1994 the three countries were collectively importing over 70 per cent of their grain requirements. Brown's argument is that China is following the same path. Its grainland per person is 0·08 ha, the same as Japan's in 1950, and the total grain-area harvest has been falling in the 1990s by over 1 million hectares a year, a rate of decline similar to that experienced by Japan, South Korea and Taiwan during their industrialization [35]. And as China's average per capita income is growing, so is its consumption of meat and other livestock products. Brown estimates that the growth in grain required for livestock feed will result in a total demand of 641 million tons by the year 2030. Given current production of about 340 million tons and no yield increase, this would generate a massive import demand, made even larger if Brown's predictions about falling grain area of 1 per cent per annum are borne out.

Needless to say, these assumptions and predictions are strongly challenged [36]. In the IFPRI model, even with exceptionally high income or population growth, China's net grain imports are not expected to exceed 96 million tons by the year 2020 [37]. Robert Paarlberg believes Brown is too pessimistic. For a variety of reasons, including evasion of land taxes and high grain-delivery requirements, Chinese farmers have kept a portion of their land unregistered. The USDA believes the true figure for the amount of arable land available is 32 per cent higher than reported, and as a consequence yields have been inflated by some 20 per cent [38]. There is thus considerably more land available for grain production than Brown reports and the published yields of 4·5 tons/ha are probably closer to 3·5 tons/ha, with some way to go before they reach the 6 tons/ha current in South Korea. An extra 1·5 tons/ha on China's actual grainland of 120 million hectares (assuming a third more land is available above the reported 90 million) could generate a further 180 million tons by the year 2020, and a decade later a further ton would bring it to Brown's estimate of demand.

So far in this book I have concentrated on cereals and, in particular, the Green Revolution cereals: rice, wheat and maize. The overall picture is somewhat better if we look at food production as a whole – not just the major staples, but other cereals and root crops, vegetables, fruits and livestock products. As incomes rise and prices fall, higher-quality grains are substituted for the lower-quality, traditional staples. However, sorghums and millets remain important in much of Sub-Saharan Africa and South Asia, and potatoes, cassava and yams are major staples in parts of Latin America, Sub-Saharan Africa and the Pacific. Starchy roots account for about one-fifth of calorie intake in Sub-Saharan Africa.

For a variety of reasons, these crops were not targeted in the early years of the Green Revolution. Rice, wheat and maize were seen as more desirable. They are sources of high-quality protein as well as carbohydrate, and are easy to store, process and prepare for eating. Most important, there was a better chance of quick and substantial yield increases. Inevitably this led to neglect of the other staples. In Indonesia emphasis on rice cultivation was at the expense of maize, taro and sago, preferred in the outer islands and often more ecologically suited to dry or swamp conditions than rice. The production of pulses (grain legumes, such as lentils, dried beans and peas) in Indonesia remained more or less unchanged during the 1960s, and in India actually fell [39]. For the developing countries as a whole, the harvest of pulses per capita has fallen over the past thirty years (Figure 7.16). This may not matter too much since, if the yields are high enough, the new cereal varieties can provide more protein per hectare than the pulses they displace. Wheat contains 8–12 per cent protein compared to the 15–25 per cent in legumes. Thus a doubling of the wheat yield could compensate for the lost legume production. But, on more marginal soils, pulses may be a cheaper and more reliable source of protein for the poor.

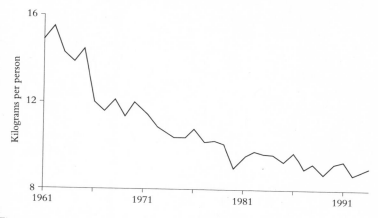

Figure 7.16 Production of pulses in the developing countries

The neglect of minor staples and legumes began to be corrected in the 1970s and 1980s following the creation of a second generation of International Agricultural Research Centres. Potatoes became the focus of the International Potato Centre in Peru (CIP), sorghums and legumes of the International Centre for Research in the Semi-Arid Tropics (ICRISAT) in India and cassava, yams, bananas, plantains and legumes of the International Institute for Tropical Agriculture (IITA) in Nigeria (see Appendix page 319 below). Production of potatoes and cassava has steadily increased, as have yields (Figure 7.17). And in recent years, new varieties of sorghum and finger millets produced in India have raised yields, although they have yet to have a significant impact in Africa [40].

As incomes rise still further, people turn to non-staple foods – vegetables, fruits, livestock products, fish and other aquatic products. Vegetable and fruit production per capita in the developing countries have increased by one-third, but the biggest percentage increases have been in livestock products, particularly milk, poultry, eggs and pigmeat (Figure 7.18). One of the biggest growth rates has been in China. At 32 kg a head, China's consumption of meat other than poultry is now higher than that of the USA, although the main component is pork rather than beef.

For the developing countries as a whole these trends have resulted in a trebling of total food production since the 1960s and an increase in per

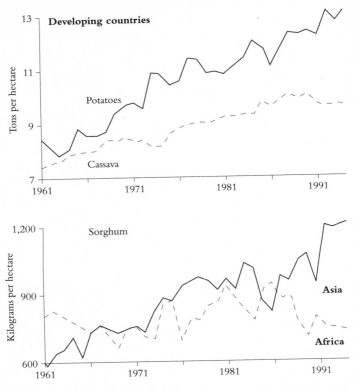

Figure 7.17 Potato, cassava and sorghum yields in the developing countries

capita food production of about a third (Figure 7.19). Most important, the rate of growth has increased in the 1980s. But there are significant regional variations (Figure 7.20). Some countries have performed much better than the average and some much worse. East Asia is the outstanding performer, accounting for a high proportion of the increase in livestock and food production generally. China's food production grew by 35 per cent in the 1980s. South Asia has also seen significant growth. Indian food output kept just ahead of population growth in the 50s, 60s and 70s, but began to accelerate in the 1980s. By contrast, there has been little

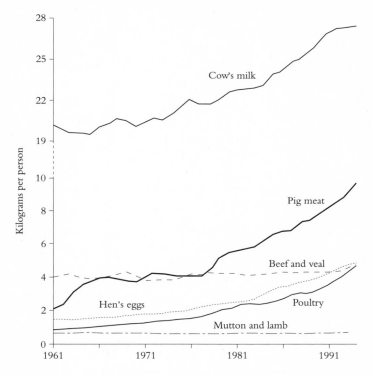

Figure 7.18 Per capita livestock production in the developing countries

improvement in Latin America and in West Asia/North Africa and Sub-Saharan Africa there has been a steady deterioration.

This review of current trends thus provides few grounds for complacency and partly justifies Lester Brown's pessimism, quoted at the beginning of the chapter. Cereal yields are showing signs of slower growth almost everywhere in the developing countries and, as a result, apart from in South Asia, per capita cereal production is levelling off or falling. The picture is better if we consider overall food production, which is rising fast in South and East Asia. However, although the trends in South Asia are upwards, they are not on track to produce sufficient food for everyone. Only in East Asia are yields and production likely to rise to the point

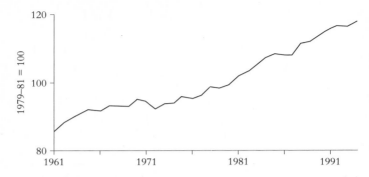

Figure 7.19 Per capita food production in the developing countries

where chronic undernutrition is banished. By contrast, the prognosis for Sub-Saharan food production is grim. Per capita cereal and food production are declining rapidly and undernutrition is likely to rise.

As I argued in Chapter 3, these trends will only be reversed if we embark on a new revolution in international agricultural research. To be successful it will have to depart in significant ways from the Green Revolution of recent decades. The development of new high-yielding varieties for the high-potential lands will continue to be vital. But it must be linked to the crucial questions of sustainability. We need urgently to know why yield growth rates are slackening in many countries and, in particular, why on some of the most intensively cropped lands there are real yield declines. Where environmental degradation is occurring we have to seek ways of reversing salinization, waterlogging, and the fall in water-tables. More sustainable production in the future will also depend on less use of pesticides and inorganic fertilizers and on reduced emissions of greenhouse gases and other global pollutants.

A return to high levels of yield growth on the Green Revolution lands will be essential if enough food is to be produced for the burgeoning urban populations of the developing countries. Equally, it will be important – but only if it is tied to the creation of greater employment and incomes – in providing food for the growing populations of rural landless and near-landless. Two hundred and eighty million of the poorest people in the developing countries live in the Green Revolution lands (see Figure 1.6).

132

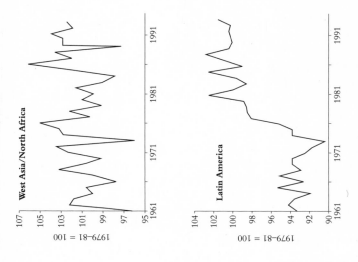

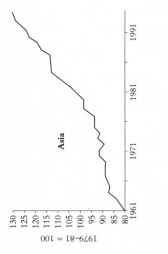

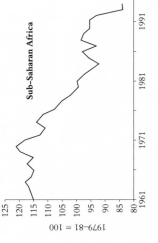

Figure 7.20 Regional per capita food production

In summary, the priorities are:

— Higher yields per hectare;
— At less cost;
— With less environmental damage;
— Creating employment and income opportunities for the landless; and coupled with
— Pricing, marketing and distribution policies that ensure that the poor gain.

Some examples of possible research themes are contained in Box 7.1.

However, the new revolution needs also to accept that these lands are now not the sole, or even main, target for innovative research and implementation. The majority of the rural poor (370 million of the poorest) live in areas that are resource-poor, highly heterogenous, and risk-prone. They inhabit the impoverished lands of north-east Brazil, the low-rainfall savannas and desert margins of the Sahel, the outer islands of the Philippines and Indonesia, the shifting deltas of Bangladesh, and the highlands of northern South Asia and the Andes of Latin America. The worst poverty

---

Box 7.1 **Examples of research themes for the high-potential lands**

— Analysis of the reasons for declines in yields of major cereals in intensively cropped, cereal-based systems

— Development of new varieties and breeds by conventional breeding and genetic engineering that deliver higher yields on the basis of lower inputs

— Construction of integrated nutrient and pest-management programmes

— Development of sustainable, high-value-export-crop systems

— Evolution of improved systems for irrigation delivery and maintenance

— Development of integrated aquaculture systems

— Reduced production of global pollutants (especially nitrous oxide and methane) from agricultural practices

— Creation of employment-generating activities based on agricultural processing and marketing

---

is often located in arid or semi-arid zones or in steep hill-slope areas that are ecologically vulnerable. There the poor are isolated in every sense. They have meagre holdings or access to land, little or no capital and few opportunities for off-farm employment. Labour demand is often seasonal and insecure. Extension services are few and far between, and research aimed specifically at their needs is sparse.

Agricultural production in such areas is limited by low rainfall and limited potential for irrigation, or steep slopes or poor soil structure, or lack of macro- or micronutrients, or presence of salts and other toxic compounds, or some combination of these. Achieving higher yields is essential but overcoming the formidable constraints requires a more integrated research approach, with a greater emphasis on improving farming systems rather than specific commodities and a higher reliance on the exploitation of resources originating within the farm and the locality. Inputs from outside, often of a high-technology provenance, will continue to be essential if even higher productivity is to be attained. But, equally, more attention will need to be paid to better use of indigenous resources, that are inherently inexpensive yet with skill and ingenuity can be used to generate higher productivity on a sustainable basis.

Future research in these areas should be aimed at:

- Higher yields per hectare;
- At very low cost;
- Making maximal use of indigenous resources, physical, biological and human;
- On a sustainable basis;

and coupled with research on

- Improving the livelihoods of rural poor households through agriculture and agriculturally related income- and employment-generating activities.

Some examples of possible research themes are listed in Box 7.2.

The complexity of these research challenges, for the high- and lower-potential lands alike, is daunting, in many respects of a greater order of sophistication that has gone before. There is certainly no case for abandoning technology. Indeed, at the outset we have to recognize there is much technology that has yet to be fully applied. In many regions, average farm yields are below those possible with only a modest increase in inputs and well below those achievable on experiment-station conditions.

Box 7.2 **Examples of research themes for the lower-potential lands**

– Improved understanding of selected critical agroecosystems, such as the highland valleys of northern South Asia

– New varieties produced through conventional breeding and genetic engineering that deliver higher yields in the face of environmental stress

– Technologies for drought- and submergence-prone rain-fed rice cultivation

– Small-scale, community-managed irrigation and water-conservation systems

– More-productive cereal-based farming systems in Eastern and Southern Africa

– Improved agro-economic systems appropriate to specific acid- and mineral-deficient soils in the savannas of Latin America

– Synergetic cropping and crop–livestock systems providing higher, more stable yields in the highlands of West Asia

– Productive and sustainable agroforestry alternatives to shifting cultivation

– Sustainable income- and employment-generating exploitation of forests, fisheries and other natural resources

---

Yet, despite what can be done with these well-tried technologies, I believe the challenge of the Doubly Green Revolution is only likely to be met by exploiting two key, recent developments in modern science. The first is the emergence of molecular and cellular biology, a discipline, with its associated technologies, which is having far-reaching consequences on our ability to understand and manipulate living organisms. The second is the development of modern ecology, an equally powerful discipline that is rapidly increasing our understanding of the structure and dynamics of agricultural and natural-resource ecosystems and providing clues to their productive and sustainable management. In the rest of this book, I argue that these and related technologies, used wisely, can help us achieve the goal of a well-fed, environmentally sustainable world.

Notes

[1] L. Brown, 1995, *Who Will Feed China? Wake-up Call for a Small Planet*, New York, NY, W. W. Norton

[2] N. Alexandratos (ed.), 1995, *World Agriculture: Towards 2010. An FAO Study*, Chichester (UK), Wiley & Sons

[3] T. Dyson, 1996, *Population and Food: Global Trends and Future Prospects*, London, Routledge

[4] P. A. Sanchez, 1983, 'Productivity of soils in rainfed farming systems: examples of long-term experiments', in IRRI, *Potential Productivity of Field Crops under Different Environments*, Los Banos (Philippines), International Rice Research Institute

[5] Alexandratos, op. cit.

[6] ibid.

[7] J. Kartasubatra, 1987, 'Indonesia', in National Research Council, *Sustainable Agriculture and the Environment in the Humid Tropics*, Washington DC, National Academy Press, pp. 393–439

[8] Dyson, op. cit.

[9] H. Linneman, J. De Hoogh, M. A. Keyser and H. D. J. Van Heemst, 1979, *MOIRA: Model of International Relations in Agriculture*, Amsterdam (Netherlands), North Holland, Report of the Project Group, Food for a Doubling World Population

[10] D. L. Plucknett, 1993, *Science and Agricultural Transformation*, Washington DC, International Food Policy Research Institute (Lecture Series)

[11] Linneman *et al.*, op. cit.; FAO, AGROSTAT.TS, Rome (Italy), Food and Agriculture Organization

[12] Alexandratos, op. cit.

[13] B. L. Bumb and C. A. Baanante, 1996, *The Role of Nitrogen Fertilizer in Sustaining Food Security and Protecting the Environment*, Washington DC, International Food Policy Research Institute (Food, Agriculture and Environment Discussion Paper 17) (from FAO Fertilizer Disk data, 1994, 1996)

[14] S. Postel, 1996, *Dividing the Waters: Food Security, Ecosystem Health, and the New Politics of Scarcity*, Washington DC, Worldwatch Institute (Worldwatch Paper 132)

[15] P. McCully (in press), *Silenced Rivers*, London, Zed Books

[16] P. Crosson and J. R. Anderson, 1992, *Resources and Global Food Prospects*, Washington DC, World Bank (Technical Paper 184)

[17] Alexandratos, op. cit.

[18] D. L. Umali, 1993, *Irrigation Induced Salinity*, Washington DC, World Bank

[19] D. Seckler, 1996, *The New Era of Water Resources Management: From 'Dry' to 'Wet' Water Savings*, Washington DC, Consultative Group on International Agricultural Research

[20] Alexandratos, op. cit.

[21] L. Brown, 1994, 'Facing food insecurity', in L. Brown (ed.), *State of the World*, New York, NY, W. W. Norton (Worldwatch Institute Report)

[22] Alexandratos, op. cit.

[23] D. O. Mitchell and M. D. Ingco, 1995, 'Global and regional food demand and supply prospects', in N. Islam (ed.), *Population and Food in the Early Twenty-first Century: Meeting Future Food Demands of an Increasing Population*, Washington DC, International Food Policy Research Institute, pp. 49–60

[24] Economic Research Service, 1992, *Agricultural Resources: Cropland, Water, and Conservation – Situation and Outlook*, Washington DC, USDA

[25] Dyson, op. cit.

[26] D. Grigg, 1993, *The World Food Problem* (2nd edn), Oxford, Blackwell

[27] P. Gypmantasiri, A. Wiboonpongse, B. Rerkasem, I. Craig, K. Rerkasem, L. Ganjanapan, M. Titayawan, M. Seetisarn, P. Thani, R. Jaisaard, S. Ongprasert, T. Radnachaless and G. R. Conway, 1980, *An Interdisciplinary Perspective of Cropping Systems in the Chiang Mai Valley: Key Questions for Research*, Chiang Mai (Thailand), Faculty of Agriculture, University of Chiang Mai

[28] P. L. Pingali, 1994, 'Technological prospects for reversing the declining trend in Asia's rice productivity', in J. R. Anderson (ed.), *Agricultural Technology: Policy Issues for the International Community*, Wallingford (UK), CAB International, pp. 384–401

[29] S. K. De Datta and K. A. Gomez, 1975, 'Changes in the soil fertility under intensive rice cropping with improved varieties', *Soil Science*, 120, 361–6

[30] J. C. Flinn, S. K. de Datta and E. Labadan, 1980, 'An analysis of long-term rice yields in a wetland soil', *Field Crops Research*, 5, 201–16

[31] N. S. Randhawa, n.d., *Some Concerns for the Future of Punjab Agriculture*, New Delhi (India) (mimeo.)

[32] Brown, 1995, op. cit.; R. Paarlberg, 1995, 'Feeding China: a confident view', Wellesley, Mass., Wellesley College (mimeo.)

[33] J. von Braun, E. Serova, H. tho Seeth and O. Melyukhina, 1996, 'Russia's food economy in transition: what do reforms mean for the long-term outlook?' *2020 Brief*, 36, Washington DC, International Food Policy Research Institute

[34] Brown, 1995, op. cit.

[35] USDA, 1994, *Production, Supply, and Demand View*, Washington DC, Economic Research Service, United States Department of Agriculture (electronic database); 1995, *World Agricultural Production*, Washington DC, Economic Research Service, United States Department of Agriculture

[36] J. Huang, S. Rozelle and M. Rosengrant, 1995, 'China and the future global

food situation', *2020 Brief*, 20, Washington DC, International Food Policy Research Institute

[37] J. Huang, S. Rozelle and M. W. Rosengrant, 1995, *Supply, Demand and China's Future Grain Deficit*, paper presented at the Workshop on Projections and Policy Implications of Medium and Long-term Rice Supply and Demand, 23–26 April 1995, Beijing (China), International Food Policy Research Institute

[38] F. Crook, 1994, 'Could China starve the world? Comments on Lester Brown's article', *Asia and Pacific Rim Agriculture and Trade Notes*, 15 September 1994, Washington DC, Economic Research Service, United States Department of Agriculture; V. Smil, 1995, 'Feeding China', *Current History*, September 1995, p. 282.

[39] J. G. Ryan and M. Asokan, 1977, 'The effects of the green revolution in wheat on the production of pulses and nutrients in India', *Indian Journal of Agricultural Economics*, 32, 8–15

[40] N. S. Jodha and R. P. Singh, 1982, 'Factors constraining growth of coarse grain crops in semi-arid tropical India', *Indian Journal of Agricultural Economics*, 37, 346–54

8 Designer Plants and Animals

> Genetic engineering will bring mishaps and stupidities in its wake.
> But, overall, it's highly likely to be a good thing, for which the
> benefits handsomely outweigh the risks.
>
> – Nicholas Schoon, *Independent* [1]

Plant and animal breeding is an art, nearly as old as agriculture itself. The
early farmers selected seed from vigorous, high-yielding plants and used
them for sowing in the following season. By degrees, wild grasses were
transformed into domestic cereals – wheat, barley, maize, rice – the process
of selection creating distinctive varieties adapted to local conditions and
needs. Farmers soon came to look for promising mutants and natural
hybrids. Bread wheat, a natural cross between emmer wheat and a wild
goat grass that arose about 7,000 years ago, somewhere to the south-west
of the Caspian Sea, was recognized and cultivated by the early farmers,
becoming the cornerstone of European and, eventually, world agriculture.
Deliberate crossing was first practised in animal breeding. Wild animals –
sheep, goats, cattle, water buffalo and pigs, among others – were tamed
and then domesticated, farmers increasingly choosing the parents of each
new generation. Much later, plant breeders learnt to cross plants and added
this to the process of selection.

  For thousands of years plant and animal breeding was a family affair,
conducted in and around the farm dwelling by men and women, using
simple ground rules and relying on intuition, their own experience and
wisdom passed on from generation to generation. Art and craft was joined
by science in the nineteenth century. Charles Darwin, in his discussions
of pigeon and dog breeds, explained the basis of selection and Gregor
Mendel, using the garden sweet pea, the particulate nature of the basis of
heredity. One consequence has been the emergence of professionals,
breeders in institutes and research stations, identifying and explaining the
underlying mechanisms and hence making the process more predictable

140

and efficient. Breeding now is considerably more sophisticated than a hundred years ago yet, in its essentials, it has little changed. Mendel, if alive today, would recognize the breeding process that led to the creation of the Green Revolution wheats and rices. The sources of dwarfing are simple dominant genes, so that crossing an imported dwarf wheat or rice with a tall indigenous variety will produce short-strawed, fertilizer-responsive offspring which nevertheless retain many if not all of the desirable characteristics of the local parent. It is a simple and powerful process, reliant for its success on the existence of naturally occurring genes capable of being easily transferred from one plant to another by the traditional methods of plant breeding.

Much of the success of plant breeding has been of this nature. Breeders have progressively improved a relatively small set of 'mainstream' varieties by crossing them with uncommon, local varieties, or in many instances wild relatives, identified as having desirable characteristics: resistance to pests or diseases, tolerance of drought, improved milling quality or flavour. Rice varieties have been bred resistant to brown planthoppers and a wide range of other pests and diseases (Table 8.1), and wheat varieties have been bred with resistance to rust and tolerance to aluminium. For the immediate future there is a pressing need, in all our important food crops, for improved resistance to viruses and insect pests, for tolerance to salt, drought and heat, for higher-quality grain and other products and for the most demanding goal of all, improved systems of nitrogen fixation.

Table 8.1 Pest and pathogen resistance in the IRRI rice varieties (++ resistant, + moderately resistant, otherwise susceptible) [2]

	Blast	Bacterial blight	Grassy stunt	Tungro	BPH 1	BPH 2	BPH 3	Green leafhopper	Stem borer
IR8								++	
IR20	+	++		+				++	+
IR26	+	++	+	+	++	++		++	+
IR36	+	++	++	++	++++			++	+
IR42	+	++	++	++	++++			++	+
IR56	+	++	++	++	++++++			++	+
IR72	+	++	++	++	++++++			++	+

Plant and, in particular, animal breeding has also been directed at improving the overall structure and physiology of crops and livestock. The aim, in simple terms, has been to increase the 'harvest index' – the proportion of the plant's or animal's energy and materials, in particular carbohydrate and protein, that goes into the final, harvested product. Beef cattle and sheep have been bred for a high 'dressing-out' percentage, the proportion of beef or lamb carcase after slaughter in the total weight; dairy cows for a high milk production and chickens for the number of eggs per day. The harvest index in cereals is the ratio of grain to straw, achieved in the early years of the Green Revolution by introducing the dwarfing genes, which had the double effect of allowing a higher uptake of nutrients and ensuring they went primarily to the grain. In recent years, the goals of plant breeders have begun to embrace the overall 'architecture' of plants in a way which is similar to animal breeders' traditional concerns that animal breeds should conform to a certain 'type' [3].

At IRRI breeders have drawn up blueprints for new rice types suited for different environments (Figure 8.1). On the well-favoured irrigated lands they are seeking very short, fast-growing varieties that can be direct seeded. The goal is a plant with only 3–4 tillers all of which produce panicles (compared with 20–25 tillers producing 15 or so panicles in existing varieties), fed by chlorophyll-rich, dark-green leaves that emerge through the panicles. Plants with this architecture should deliver a very high harvest index and a maximum yield of 15 tons/ha. The breeders have already gone a long way towards this goal. In 1994, Gurdev Khush and his team at IRRI produced a new 'super-rice', with about 8 tillers, all bearing panicles each with about 200 grains (compared with 100 on existing varieties). Its yield potential, once pest and disease resistance has been added, is about 13 tons. For the rain-fed lowlands, they are looking for more medium-sized varieties, tolerant of submergence and deep-rooted to improve their resistance to drought. Perhaps the most imaginative target is for the uplands, where they envisage a perennial rice that has the ability to fix its own nitrogen. These goals can be partly achieved by transferring individual genes conferring particular traits, but much of the strategy depends on manipulating large numbers of interacting characteristics. The new 'super rice' was produced as a result of a thorough search of IRRI's rice collection for plants with fewer tillers, more grains per panicle and stronger roots.

Sometimes conventional plant breeding, using sophisticated techniques,

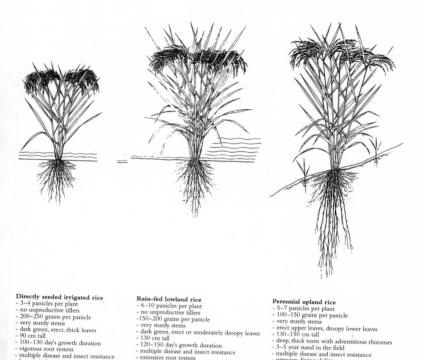

Directly seeded irrigated rice
- 3–4 panicles per plant
- no unproductive tillers
- 200–250 grains per panicle
- very sturdy stems
- dark green, erect, thick leaves
- 90 cm tall
- 100–130 day's growth duration
- vigorous root system
- multiple disease and insect resistance
- harvest index 0·6
- 13–15 t/ha yield potential

Rain-fed lowland rice
- 6–10 panicles per plant
- no unproductive tillers
- 150–200 grains per panicle
- very sturdy stems
- dark green, erect or moderately droopy leaves
- 130 cm tall
- 120–150 day's growth duration
- multiple disease and insect resistance
- extensive root system
- strong submergence tolerance
- strong grain dormancy
- 5–7 t/ha yield potential

Perennial upland rice
- 5–7 panicles per plant
- 100–150 grains per panicle
- very sturdy stems
- erect upper leaves, droopy lower leaves
- 130–150 cm tall
- deep, thick roots with adventitious rhizomes
- 3–5 year stand in the field
- multiple disease and insect resistance
- nitrogen-fixing ability
- aromatic grain
- 3–4 t/ha yield potential

Figure 8.1 The goals for rice breeding [4]

can produce striking results. There is currently an attempt to produce a new bread wheat by crossing the goat grass *Triticum tauschii* with durum wheat. The goat grass has only 14 pairs of chromosomes, while durum has 28. However, chemical treatment can be used to double the goat grass complement, so making crossing viable. The hope is that such crosses will improve resistance to such diseases as helminthosporium, fusarium and Karnal bunt, and possibly tolerance of salt and drought.

Some of the most exciting developments in conventional plant breeding have been the production of hybrid maize and rice (Chapter 4). The search is now under way for inexpensive, large-scale means of producing hybrid wheat. The advantage of these crosses is their enhanced vigour, which typically increases yields 20 per cent or more over the parents. The disadvantage is that farmers, if they are to maintain yields, have to buy their seed each year. One goal of the breeders is to develop a maize and a rice that will reproduce asexually – through apomixis. Then the seed of a hybrid with this trait can be retained from one year to the next. In the case of maize it is possible, although difficult, to transfer apomixis from a close relative, *Tripsacum*. This and the preceding examples illustrate the crucial importance of maintaining germ plasm representing the wide range of variation in crop plants and their relatives. FAO estimates there are globally some 6 million accessions stored in more than one thousand gene banks. Some of the biggest collections in the developing countries are in China and India, and there are major collections in most of the International Agricultural Research Centres [5].

Conventional breeding techniques such as these have much to offer but, as has been long recognized, they have practical limitations. The process of crossing two parents, each with desirable characteristics, in the expectation of producing offspring with a new and improved combination of these characteristics is essentially a random process. There is normally a degree of success, and repeated crossing and selection will, with care and attention, result in superior varieties. Inevitably, though, there are trade-offs. While some desirable characteristics may come together, others may be lost. Potential yield may increase, but often at the expense of pest or disease resistance or some other characteristic which is already present, such as superior grain quality. And as crops and livestock become more highly bred, more sophisticated in genetic terms, progress becomes more difficult because of the complexity of genetic manipulation that is required. There are also natural limitations to conventional plant breeding. Traits which a breeder wishes to incorporate in a plant or an animal may not be present in any species with which a cross can be made, although they may occur in quite unrelated species.

Conventional plant breeding is also a relatively slow process. In tropical climates, using irrigation, it is possible to complete three generations of a cereal, such as rice, in a year. But in many important crops, generation turnover is a great deal slower. For some of the major cash crops of the developing countries – rubber, oil palms, cacao, coffee and coconuts – it

144

takes decades to achieve yield advances which in rice would be realized in at most two to three years. Plant breeders are accustomed to this pace of advance, but for agronomists and farmers it can be frustrating.

It is for these reasons that the revolutionary achievements of cellular and molecular biology are so important [6]. They are based on remarkable advances in laboratory techniques which are enabling scientists to probe and experiment with the processes that are fundamental to life, increasing our understanding and, at the same time, making it possible for us to manipulate these processes to our advantage. Under the general title of biotechnology, they are already having an enormous impact on medicine and we are just beginning to realize their potential in plant and animal breeding (Box 8.1).

Box 8.1 Potential applications of biotechnology in agriculture [7]

Crop improvement
- Protoplast fusion and somatic hybridization to produce new crosses
- Disease-free plant propagation
- Production of genetic maps
- Biological nitrogen fixation
- Genetically engineered male sterility, to produce hybrid varieties
- Transgenic plants for pest resistance
- *In vitro* germ plasm conservation, storage and distribution

Livestock improvement
- Production of growth hormones using engineered bacteria
- Embryo manipulation to introduce new traits
- Transgenic animals for better feed efficiency
- New vaccines
- Disease diagnosis

From cellular biology the most important practical contribution has been the ability to isolate plant and animal cells and grow or raise them into fully mature, whole organisms, capable of independent reproduction. In plants the techniques go under the name of cell or tissue culture. Sometimes it is an embryo that is cultured. In 1977 IRRI successfully transferred resistance to grassy stunt virus from a wild rice, *Oryza nivara*, using this technique [8]. Wild rice has a different number of chromosomes; crosses with domestic rices are possible but the embryos will not normally survive. However, they can be 'rescued' soon after fertilization, cultured on a medium in a test tube until they germinate and then planted out in a greenhouse or in the field. Another possible cross with the wild rice *Oryza australiensis* could be a source of drought tolerance.

Tissue culture also allows for a speeding up of the process of plant breeding [9]. Self-fertilizing plants, such as wheat and rice, are normally 'true-breeding' – since the chromosomes in each pair are identical, each offspring is the same as its parent. When a cross is made between two different parents the chromosomes are no longer all identical and it takes many generations of self-fertilizing before they become so. However, it is possible to fast-track the process by culturing the anthers – the male reproductive organs – of the cross. The cells in the anthers are haploid, containing only one member of each chromosome. They can be placed in a culture where they double each chromosome, so producing a young plant which will now produce identical offspring containing the new characteristic. Breeders in China have produced new varieties in only five years using this technique, compared with the twelve or so years of the normal cycle.

Cell culture involves removing the wall from an individual cell, by means of a mixture of digestive enzymes, leaving it contained in a flexible membrane. This new stripped cell, known as a protoplast, can then be cultured on a medium rich in sugars, vitamins and minerals. Under the influence of an optimum combination of plant hormones the cell divides, eventually producing an embryo which can be encased in a matrix, dried and subsequently planted as if it were a seed. The resulting plant is a product of the original cell alone and hence faithfully carries its characteristics. It is a valuable technique for making exotic crosses [10]. Where different plants may not cross naturally or under conventional plant-breeding techniques, it is relatively easy to take cells and convert them to protoplasts, which can then be fused. So far the technique has worked with plants that are relatively closely related, for example, potatoes and tomatoes, or different brassicas, such as cauliflower and cabbage. Protoplasts from tomatoes with resistance to bacterial wilt or soft rot are irradiated with gamma rays to break up their chromosomes into small pieces. They are then fused with potato protoplasts. The new protoplasts are cultured, producing many thousands of plants which are screened for the resistance, those showing the trait being multiplied in conventional ways.

It turns out that cells in such cultures also exhibit a high frequency of mutations: in particular so-called silent genes are activated. This has been used to produce new lines of tomatoes and bananas resistant to disease, and higher-yielding potatoes. The rapid cell cycling in cultures also results in chromosome breakage and recombination, and this is being exploited to develop resistance to a common virus which attacks wheat and other cereals, known as barley yellow dwarf virus (BYDV) [11]. Resistance is present in wild grasses such as intermediate wheatgrass. These will cross with wheat but the chromosomes will not combine. However, by taking cells from the crosses and putting them into culture it has been possible to raise a few plants in which the breakage and recombination in the culture have resulted in the virus–resistance genes becoming part of the wheat chromosome.

Conventional livestock breeding, like plant breeding, is limited by the random nature of normal crossing and also by long generation times. Through techniques that have some similarity with those of plant-tissue culture, advances in cellular biology and animal physiology have made it possible to focus and speed up the breeding process. The most commonly used technique is to take unfertilized eggs from their mothers, fertilize the eggs in the laboratory and then re-implant them. Normally a cow gives birth to four calves in a lifetime, but will produce thousands of eggs. By collecting the eggs, fertilizing and returning them it is possible for superior breeds of cow to give birth to over 100 calves in a lifetime. The technique also permits the embryos to be manipulated before they are returned. It is possible to divide embryos, so increasing the incidence of twinning, and hence the rate of multiplication of new breeds.

Cellular techniques on their own can greatly increase the power of conventional plant breeding, but in combination with the revolutionary advances in molecular biology they open a new world in which breeders can deliberately design and engineer new plant and animal types, speedily and with much less reliance on random processes. This revolution in genetic engineering has its origins in the discovery by James Watson and Francis Crick, just over forty years ago, of the structure of the DNA molecule which contains the genes of living organisms. As they brilliantly showed, the molecule consists of two chains wrapped around each other in a 'double helix'. Along the chains are sequences of four chemical bases in various permutations which constitute the genetic code – an alphabet which, despite relying on only four letters, delivers messages of great subtlety. Subsequent research has shown how living organisms read the code, translating the sequence of bases in the DNA into proteins, which then work as enzymes, antibodies and hormones, building and maintaining the tissues that go to make up whole plants and animals.

Each gene consists of many thousands of bases linked across the two strands of the DNA helix. Lengths of DNA, surrounded by a shell of protein, constitute the chromosomes, which are usually present as pairs in the nuclei of plant and animal cells. There are 21 paired chromosomes in wheat, 12 in rice, 30 in cattle and 27 in sheep. The chromosomes and the genes they contain constitute an organism's genome – a kind of encyclopedia containing thousands of recipes. On this analogy, each pair of chromosomes is an individual volume of the encyclopedia, whose individual entries, the 1,000 or so genes, each provide a recipe for making a protein. A gene may act alone, producing an individual protein whose specific function creates a recognizable characteristic in the plant or animal – resistance to a pest, for example – or a gene may combine with other genes in complex ways. Yield, for instance, is rarely the function of the actions of a single gene.

At one time, mapping a genome, identifying the sequence of bases in each strand of DNA, seemed to be an impossible task. New laboratory techniques which enable pieces of DNA to be recognized and characterized have made the task easier, although it requires time and patience. The genome of yeast has now been fully mapped and the Human Genome project, a worldwide collaboration which has set itself the task of mapping the 50,000–100,000 genes in the human genome, has so far identified nearly 17,000 genes [12]. Fortunately, there is considerable similarity in the chromosomal layout among related species. Although wheat has 21 very large chromosomes and rice 12 small ones, their genes and, indeed, the genes of all cereals are remarkably similar. They are also arranged in the same sequences along the chromosomes. The difference between wheat and rice results from the long repetitive sequences of 'junk DNA' in the wheat chromosomes, which has no detectable purpose. Because the genome of rice is smaller it is easier to work with. The Japanese Rice Genome Project is now halfway through its task of mapping rice's 30,000 or so genes.

The most important laboratory tool in genome mapping is a class of compounds known as restriction enzymes, which have been developed by bacteria, apparently as a defence against invasion by alien DNA. They act as molecular scalpels. About 1,000 restriction enzymes are known.

Each is capable of recognizing a particular sequence of bases in the DNA helix and when it does so, the enzyme makes a cut. By using combinations of these enzymes it is possible to divide the DNA up into pieces of about the length of a gene (Figure 8.2). The next step is to incorporate each length in a bacterium, either via a form of virus (a bacteriophage or phage) or as a plasmid, a simple coil of DNA. The bacterium, most commonly a harmless form of *Escherichia coli* (*E. coli*), is then cultured on an agar medium. The next stage is the most difficult, identifying the presence of individual whole genes in the bacterial colonies, using a DNA probe, but once a gene has been identified the carrying plasmid or phage can be multiplied many times to produce the gene in large quantities. It is a painstaking, often frustrating, piece of detective work.

When the DNA helix is cut, a pair of bases are left exposed at the ends. They can be joined to another pair at the cut ends of a second piece of DNA, this time with the help of an enzyme known as DNA ligase, which acts as a molecular suture. By this process of microsurgery, genes can be taken from one DNA helix and spliced into another and hence transferred between chromosomes. Although an artificial act it is, in essence, the same process as occurs when the plant or animal breeder crosses one plant with another, or one animal with another. During the crossing process, chromosomes transfer pieces from one to another, but the new combinations are usually randomly determined. The great advantage of recombinant DNA technology – or genetic engineering, as it is popularly called – is that the new combinations are determined beforehand and, with skill and care, are precisely achieved. As a result, the plant breeder is no longer restricted to the genetic variation that arises in traditional breeding programmes.

Of course, insertion of a gene into a new host environment is not straightforward. Moving genes from one species to another unrelated species, for example from a pea plant to a wheat plant, or more radically from an animal to a plant species, requires various supportive measures if the gene is not to be rejected. Genes have control regions – promoter and terminator sequences – which determine how a gene is made and its response to environmental signals. Genetic engineers take these regions from a gene that works well in the new species and then attach them to the genes they wish to introduce. For example, a gene producing egg white (albumin) in chickens can be introduced into a clover plant by adding control sections from a successful clover gene. The result is a clover

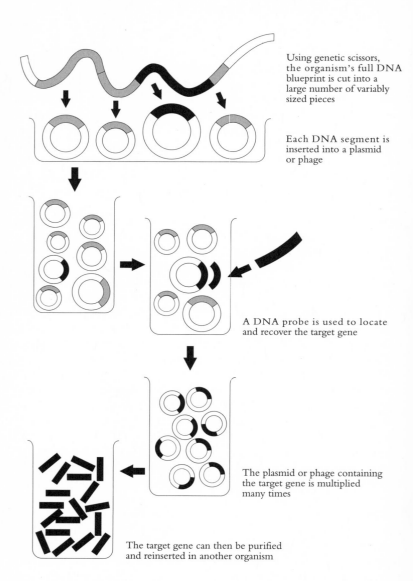

Using genetic scissors, the organism's full DNA blueprint is cut into a large number of variably sized pieces

Each DNA segment is inserted into a plasmid or phage

A DNA probe is used to locate and recover the target gene

The plasmid or phage containing the target gene is multiplied many times

The target gene can then be purified and reinserted in another organism

Figure 8.2 Isolating genes [13]

plant that accumulates albumin in its leaves, making it of greater value to grazing animals. Genes can also be made more effective by adding a stronger promoter and then returning them to their original host. Using this technique, livestock have been engineered to produce greater quantities of bovine growth hormone. Most radical of all is the potential to create entire artificial chromosomes and insert them into new organisms.

The process of transferring a gene to a new plant or animal can follow one of several routes. One of the earliest techniques was to colonize the gene in *Agrobacterium tumefaciens*, a bacterium that invades potatoes, tomatoes and lucerne (alfalfa), among other plants. It is a surprisingly easy process: pieces of leaves are dipped into a suspension of bacteria, then cultured to produce whole plants containing the new gene. For wheat, barley and rice, which are not invaded by this bacterium, the gene DNA has to be injected more forcefully. One approach is to use a high-voltage electric pulse which permits DNA to enter a protoplast; alternatively the DNA is applied as coat to inert tungsten or gold particles which are then fired into the plant cell with a microparticle gene gun. In 1996 the British company Zeneca discovered an extraordinarily simple, but effective, method: if plant cells, DNA and crystals of silicon carbide are shaken together in a test tube, the crystals punch holes in the cell walls, allowing the DNA to enter [14]. It is customary for millions of cells to be treated; those in which the new gene have lodged are identified by means of a marker and cultured.

For animals the technique is to inject the DNA into the nucleus of a single cell of a fertilized egg using a very-fine-bore glass needle, finer than a human hair. This is usually done 12–14 hours after fertilization. The egg is then implanted in the uterus of a mature female and raised in the normal way. If these transfer processes are successful, the resulting transgenic plants and animals will not only carry the new gene and exhibit its properties but, on maturity, will be capable of passing it on to their progeny through the normal process of sexual reproduction.

Genetic engineering has a special value for agricultural production in developing countries. It has the potential to address the specific problems detailed in the previous chapter, creating new plant varieties and animal breeds that not only deliver higher yields but contain the internal solutions to biotic and abiotic challenges, reducing the need for chemical inputs such as fungicides and pesticides, and increasing tolerance to drought, salinity, chemical toxicity and other adverse circumstances. Most important, genetic engineering is likely to be as valuable a tool for the lower-potential lands as for the high-potential. It can be aimed not only at increasing productivity but at achieving higher levels of stability and sustainability. Much depends on how the power of the technique is used, what targets are set, how the process is managed and funded and how the products are used.

Box 8.2 lays out the essential steps in a genetic-engineering programme, using as an example the objective of improving pest resistance in a crop plant. Some of the most promising applications lie in the conferment of resistance to pests and to bacterial and virus diseases. Resistance to viruses can be achieved by transferring to plants certain genes that encode for the virus coat protein. When the protein is released by the plant, it prevents the invading virus from shedding its coat. The DNA virus cannot escape into the plant cell and hence is unable to replicate. The technique has been used against several important viruses: rice stripe virus, lucerne mosaic virus and potato leaf roll virus. Another approach is to use ribozymes, so-called 'gene shears'. These are a form of RNA (a close relative of DNA) that are present in certain parasites of viruses, which can target and destroy another piece of RNA. Certain genes encode for the production of gene shears and, introduced into plant or animal cells, they will destroy invading viruses.

Genetic engineering offers a number of novel ways of combating pests. Bacteria sometimes produce proteins that are insecticidal. In the insect gut they break down the cells necessary for nutrient absorption: the insects stop feeding and die. One bacterium, *Bacillus thuringiensis* (*Bt*), produces crystalline proteins of this kind that kill many pests, particularly important caterpillar pests, but are relatively safe to beneficial insects such as honey bees and insect parasites and predators. Also, they do not harm humans. The genes encoding for these proteins were first isolated in the early 1980s and have now been transferred to a number of crops (Figure 8.3). Farmers in the USA are regularly growing maize, potatoes and cotton containing the *Bt* gene. It has also been transferred to rice, where its potential lies in the control of stem borers, an important pest against which conventional plant breeding has made little headway [15].

The *Bt* gene has a particularly important role to play in integrated pest management (IPM) programmes (see Chapter 11). Although the toxin is lethal to stem borers, it has little effect on the other pests of rice. However, there are drawbacks. Just as insects become resistant to synthetic chemical

---

Box 8.2 **The steps in a genetic-engineering programme**

- Determine the objective, e.g. resistance to a pest in a particular crop plant
- Identify the possible mechanism of resistance and the proteins involved
- Identify the likely source of a useful gene, e.g. in a wild plant
- Culture DNA fragments from the source
- Find or synthesize a probe for the appropriate gene
- Isolate the gene
- Decode (sequence) the gene to determine its structure
- Redesign the gene for the new plant environment
- Insert the gene into single cells of the target plant
- Grow the transformed cells into complete plants
- Test the new, transgenic plants for their resistance and other characteristics in the laboratory and in the field
- Thoroughly assess likely hazards, e.g. harm to beneficial insects
- Multiply the new crop plants for distribution to farmers
- Distribute the new plants as part of a new programme of integrated pest management

---

insecticides applied to crops, so they are likely to evolve resistance to insecticides produced by the plants themselves. There are currently eight species of insect in the USA resistant to *Bt* toxins, and there is a danger that widespread use of engineered plants containing the *Bt* gene will increase the general level of exposure to the toxin among pest populations, so hastening the evolution of resistance [16]. One approach to slowing this process may be to mix plants, with and without the *Bt* gene, either in an individual field or by ensuring that farmers rotate the resistant and non-resistant varieties [17]. In this way the evolutionary pressure is lessened.

In the long run, however, the best approach is to develop and utilize as many possible forms of resistance as possible, including the introduction into plants of multiple toxic genes, each with a different mode of action. A promising companion to the *Bt* gene is another gene that encodes for a proteinase inhibitor contained in the tropical giant taro and confers high resistance to insect attack. The protein probably inactivates the proteases of an insect, so starving it to death, a very different insecticidal mechanism from that of the *Bt*. Another potentially useful gene makes the polyhedrosis

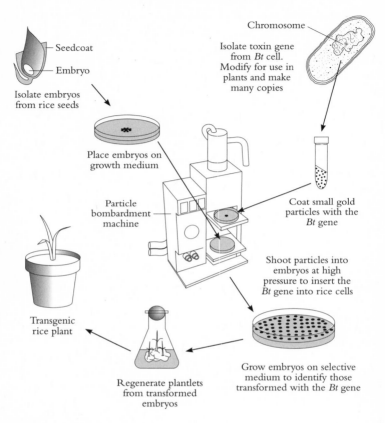

Figure 8.3 Steps in the production of *Bt* rice [18]

baculovirus (NPV) more lethal to boll-worms. Since the toxin is only expressed in cells where the virus is actively growing there is little or no danger to non-target organisms.

In the developed countries, much of the genetic engineers' attention has been focused on the creation of herbicide tolerance. The general objective is to make it safe for herbicides to be applied to growing crops, so that the weeds are killed but the crops are left unharmed. Several techniques are currently being developed. For example, if a herbicide works by inhibiting a particular enzyme, the encoding gene can be engineered for stronger expression, perhaps as much as ten times the tolerance level. This has been achieved for crops such as maize and soybeans treated with the herbicide glyphosate. It is also possible to use the genes in soil bacteria that metabolize herbicides, for example 2,4-D and bromoxynil, into innocuous chemicals.

However, there are understandable fears that engineering greater herbicide tolerance may encourage the indiscriminate use of herbicides. Because hand weeding in the developing countries is an important source of rural employment, greater use of herbicides could reduce income–earning opportunities for the poor (see Chapter 5). Nevertheless, there may be advantages in certain situations. Where herbicides are already in use, the ability to spray in the growing crop with so-called post–emergence herbicides, to which the crop is resistant, will reduce heavy dosing with precautionary herbicides before the crop emerges. Herbicides could become better targeted, only employed when necessary, hence reducing the levels of environmental contamination. There is also a special role for post–emergence herbicides on certain tropical soils in need of better soil conservation. Soil can be protected from erosion by sowing seed direct into the stubble of the previous crop. However, this can result in severe weed problems. If hand weeding is not available, post–emergence herbicides will be necessary.

Identifying and transferring genes for resistance of various kinds is now relatively easy. Far more difficult is the genetic engineering of nitrogen fixation [19]. Nitrogen-fixing bacteria convert atmospheric ammonia to nitrogen using an enzyme, nitrogenase. To accomplish this they need energy, which the symbiotic forms, known as rhizobia, obtain by living inside plants. The plants furnish some of the products of their photosynthesis and the bacteria reciprocate by supplying nitrogen. Virtually all symbiotic relations are with legumes, such as peas, lupins, clover and lucerne (the only exception is the tropical plant, *Parasponia*). Cereals and all other non-leguminous food crops have to rely on other sources of nitrogen and, increasingly, on the inorganic nitrogen fertilizer purchased by the farmer.

The first task is to try to improve the efficiency of symbiotic bacteria, and this can be done by transferring genes from one strain to another. More challenging is the aim of creating symbioses with wheat, rice and other cereals. In the laboratory it has proved possible, by chemical treatment, to permit rhizobia to invade wheat and rice plants, which then develop nodule-like structures, but scientists are still a long way from turning this into a stable, heritable association. The biggest challenge is to get plants to fix their own nitrogen. In 1971 Ray Dixon of the Agricultural Research Council's Nitrogen Fixation Unit at Sussex University in the UK successfully transferred the complex of genes known as *nif* from a free-living, nitrogen-fixing bacterium, *Klebsiella*, to the bacterium *E. coli*. Since then the *nif* cluster in *Klebsiella* has been identified as consisting of twenty genes in a row which code for the manufacture of nitrogenase and ensure it works efficiently. In theory, we could transfer the whole cluster to normal plant cells. Unfortunately, as John Postgate puts it, 'bacterial genes do not make sense to plants (nor to animals for that matter – my pet dream is of a nitrogen-fixing goat that could convert all our junk paper and cellulose waste direct to meat)' [20]. Although plants, animals and bacteria share the same genetic code, they have different reading systems: the promoters, the number of genes read at a time and the ways of handling the genetic message all differ. These are formidable obstacles and will take years of fundamental research before they are overcome. Yet if this goal can be achieved the benefits would be incalculable. The environmental pollution from inorganic fertilizers would be greatly reduced, as would the costs to poor farmers of producing higher cereal and other crop yields on marginal lands.

One way livestock can benefit from genetic engineering is through the provision of better feed. Pigs and chickens need lysine in their diets but the content in cereals is poor compared with that in legumes, such as peas and lupins. On the other hand, legumes are poor in the sulphur amino acids, methionine and cysteine, necessary for cattle and sheep if meat, milk and wool productivity is to be increased. High levels of these amino acids are present in sunflower seeds and the chicken–egg protein, ovalbumin, and the encoding genes have now been cloned and inserted into peas. The next step is to insert the same or similar genes in forage legumes, such as lucerne and clover, with suitable promoters so that they are expressed in the foliage. An alternative approach is to introduce genes for sulphur amino acid biosynthesis, present in the bacterium *E. coli*, direct into sheep, bypassing the need for improved fodder.

Most of the fibre in livestock fodder is not digested, and tropical forages have particularly high fibre and lignin contents. Poor digestibility results in lower food intake and slow lightweight gain. One approach is to use gene shears to block the pathway which produces lignin. So-called 'brown midrib' mutants with this effect sometimes occur in maize, sorghum and millet. Fifty per cent less lignin is produced and the digestibility is 10–30 per cent higher. Research is now aimed at cloning these genes and inserting them into forage legumes.

As with plants, some of the most promising applications of genetic engineering in animals are directed at pest and disease control. Considerable progress has already been made in the production of genetically engineered vaccines. Hitherto vaccines have consisted of dead or live, but attenuated, viruses or other pathogens. The pathogens in this form are harmless but they provide the antigens that stimulate production of antibodies in the vaccinated animal and hence give it protection if invaded by the live, virulent pathogen. However, this traditional method of vaccination can be hazardous. Live vaccines, in particular, carry material other than the antigens which may cause undesirable side-effects. The advantage of genetically engineered vaccines – produced by cultures of bacteria which contain antigen-encoding genes – is that they consist solely of the required antigens. Genes encoding for resistance to pests and diseases can also be directly transferred from one animal to another or even, in some cases, from plants to animals. Blowflies that burrow into the skin of livestock animals are killed by chitinase, the enzyme which breaks down the chitin of which their exoskeleton is composed. Genes encoding for chitinase are found in various plants. Another possibility is to introduce the *Bt* gene into livestock.

The potentials for genetic engineering are almost endless. But there are serious hazards, some easily perceived, others yet to become apparent [21]. I have already referred to the likely increased use of herbicides following the introduction of herbicide-resistant crops and to the evolution of resistance to engineered toxins, such as *Bt*. Perhaps the most obvious hazard is the possibility of a transferred gene being further passed through natural processes to another organism, with detrimental effects [22]. A recent controversy, attracting much media attention, has arisen from the importation to Europe of US maize engineered for resistance to borers. The hazard lies less in the transferred resistance gene, and more in the associated marker gene, one resistant to an antibiotic, used to identify whether the resistance gene is present. The maize is primarily intended as animal feed and, as opponents of the importation argue, with some justification, there is a danger of the antibiotic resistant gene transferring to the cattle who eat the maize.

Many crops have wild relatives and hybrids may naturally occur, which would permit the new genes to pass out of the crop. In the USA, sorghum crosses with Johnson grass, producing hybrids that are aggressively weedy. Rice plants are self-pollinating, but not exclusively so. A degree of cross-pollination does occur under natural conditions, both among cultivated rice and between cultivated and wild rices. There is thus a likelihood in Asia of *Bt* transferring to the wild relatives, particularly *Oryza nivara* and *O. rufipogon* [23]. These could then become serious weeds, although the evidence suggests that they are not presently limited by stem borers or other caterpillars. Weediness is a complex phenomenon resulting from a combination of many characteristics. The transfer of genes which increase a plant's competitiveness or resistance to stress will make it more likely to become a weed. In some cases this may be the result of unforeseen interactions between introduced genes and those already present.

Another hazard lies in the possibility of the use of virus resistance genes, such as those which encode for the virus coat protein, resulting in the evolution of new strains of viruses or an increase in the range of crops they attack [24]. There is evidence that exchanges may occur between the nucleic acid of a coat protein gene and the nucleic acid of a related virus, changing the characteristics of the virus. Another possibility is the coat protein produced by the gene enclosing another virus, which makes it more easily transmittable to other crops. As this example shows, we are increasingly dealing, in the words of Jane Rissler and Margaret Mellon, of the US Union of Concerned Scientists, with risks that 'may be missed simply because the understanding of physiology, genetics, and evolution, among other disciplines, is limited. What such risks might be are, by definition, hard to imagine' [25].

The developed countries are clearly better equipped to assess these hazards. They can call on a wide range of expertise and most have now set up regulatory bodies and are insisting on closely monitored trials to try to identify the likely risks before genetically engineered crops and livestock are released to the environment. So far, few developing countries have put such regulation in place, raising fears that developed-country corporations may use developing-country sites as unmonitored laboratories, with potentially severe consequences. My personal belief is that the hazards are often overstated, but if the evident benefits are to be realized in the developing countries then developed-country scientists in both public agencies and private industry must ensure the hazard assessments are as rigorous as in their own countries.

More important than the potential hazards, at least to my mind, is the question of who benefits from genetic engineering, and indeed from conventional breeding processes. If the work is privately funded, the products may be expensive and protected by highly restrictive patents. Modern plant and animal breeding, whether conventional or reliant on genetic engineering, is often very expensive. According to one estimate, it costs about $1 million to clone a single gene [26]. Nevertheless, the market opportunities and the potential returns are considerable. In the USA over 120 small companies are specializing in agricultural biotechnology, spending an average $4–$6 million a year on research and development [27]. Many large corporations are also involved, often marketing their own seed and, in the case of herbicide-resistant crops, selling the seed in conjunction with the herbicide.

Investment is dependent on private companies realizing an adequate return in commercial terms – covering their development costs, a profit for shareholders and funds to invest in new laboratories and new products. The existence of patent law is thus a crucial incentive but, in some instances, companies have made very comprehensive patent claims. Agracetus, a US company which was the first to transfer genes in soya beans using the 'gun' technique, has been granted patent rights in Europe to all genetically engineered soya beans, regardless of the method of transfer employed. Clearly, if vigorously applied, this kind of patent could limit independent research and development.

Genetic engineering is a highly competitive business and inevitably, the focus of biotechnology companies has been on developed-country markets where potential sales are large, patents are well protected and the risks are lower. In this situation public–private partnerships are going to be essential if developing countries are to benefit [28]. The Rockefeller Foundation is spending some $5–$6 million a year on rice biotechnology, principally in Asia, with the aim of supporting developing-country laboratories [29]. The International Agricultural Research Centres have also become increasingly involved and are pioneering partnerships which aim to transfer the techniques and achievements of developed-country biotechnology. One project involves CIAT in Colombia, the University of Wisconsin and Agracetus, a Wisconsin-based company, in the development of beans resistant to bean golden mosaic virus. The process of gene transfer remains a secret of Agracetus, but the resulting genetic material is made available to CIAT, who will pass it to public-sector seed suppliers. So far, the agreements covering genetic engineering in rice have also proven reasonably favourable for the developing countries. IRRI's work on the transfer of *Bt* to rice is using genes from two private companies [30]. In one case, IRRI has paid a fee to Plantech of Japan to use the gene for research purposes and has an option to buy the gene outright. In the other case, the gene has been provided free by Ciba–Geigy of Switzerland. *Bt* rice developed by IRRI can be made freely available through the developing world, but not to Australia, Canada, Japan, New Zealand, the United States and the members of the European Patent Convention. In this way developing countries are able to benefit, while Ciba–Geigy protects its patent rights in the developed countries.

Through genetic engineering we have the potential to develop crops and livestock that are resistant to pests and diseases, that can compensate for mineral deficiencies and withstand salinity, toxins and drought, and can make more efficient use of sunlight, water and nutrients. The potential is enormous, but there are risks that are not insignificant. As Nicholas Schoon remarks in the quotation at the beginning of this chapter, there are bound to be mistakes, but they can be minimized by employing the technology wisely, in the context of sophisticated ecological and physiological knowledge.

Notes

[1] N. Schoon, 1996, 'Nothing to fear from techno-corn'. *Independent* (London), 11 December

[2] G. S. Khush, 1990, 'Multiple disease and insect resistance for increased yield stability in rice', in IRRI, *Progress in Irrigated Rice Research*, Los Banos (Philippines), International Rice Research Institute; 1992, 'Selecting rice for simply inherited resistances', in H. T. Stalker and J. P. Murphy (eds.), *Plant Breeding in the 1990s: Proceedings of the Symposium on Plant Breeding in the 1990s*, Wallingford (UK), CAB International, pp. 303–22

[3] M. Lerner and H. P. Donald, 1966, *Modern Developments in Animal Breeding*, London, Academic Press

[4] IRRI, 1989, *IRRI Toward 2000 and Beyond*, Los Banos (Philippines), International Rice Research Institute

[5] FAO, 1996, *Seeds of Life*, World Food Summit, Rome, Food and Agriculture Organization; T. T. Chang, 1992, 'Availability of plant germplasm for use in crop improvement', in Stalker and Murphy, op. cit., pp. 17–35

[6] Stalker and Murphy, op. cit.

[7] R. W. Herdt, 1991, 'Perspectives on agricultural biotechnology research for small countries', *Journal of Agricultural Economics*, 42, 298–308

[8] IRRI, 1993, 'The tools of rice biotechnology', *IRRI Reporter*, March 1993, 3–9

[9] P. J. Larkin (ed.), 1994, *Genes at Work: Biotechnology*, Canberra (Australia), CSIRO

[10] ibid.

[11] ibid.

[12] G. D. Schuler *et al.*, 1996, 'A gene map of the human genome', *Science*, 274, 540–46; A. Goffeau *et al.*, 1996, 'Life with 6000 genes', *Science*, 274, 546–8

[13] Larkin, op. cit.

[14] C. Cookson, 1994, 'The nature of things: evolution of long-running cereals', *Financial Times*, 2, 4 July, XIII

[15] IRRI, 1996, *Bt Rice: Research and Policy Issues*, Los Banos (Philippines), International Institute for Rice Research (IRRI Information Series No. 5)

[16] B. E. Tabashnik, 1994, 'Evolution of resistance to *Bacillus thuringiensis*', *Annual Review of Entomology*, 39, 47–79

[17] IRRI, 1996, op. cit.

[18] W. McGaughey and M. Whalon, 1992, 'Managing insect resistance to *Bacillus thuringiensis* toxins', *Science*, 258, 1451–5; J. Rissler and M. Mellon, 1996, *The Ecological Risks of Engineered Crops*, Cambridge, Mass., MIT Press

[19] J. Postgate, 1990, 'Fixing the nitrogen fixers', *New Scientist*, 3 February, pp. 57–61

[20] ibid.

[21] H. A. Mooney and G. Berndardi (eds.), 1990, *Introduction of Genetically Modified Organisms into the Environment*, Scientific Committee on Problems of the Environment, Chichester (UK), Wiley & Sons; R. Casper and J. Landsmann (eds.), 1992, *The Biosafety Results of Field Tests of Genetically Modified Plants and Microorganisms*, Braunschweig (Germany), Biologische Bundesanstalt für Land- und Forstwirtschaft

[22] Rissler and Mellon, op. cit.

[23] M. T. Clegg, L. V. Giddings, C. S. Lewis and J. H. Barton (eds.), 1993, *Report of the International Consultation on Rice Biosafety in Southeast Asia, 1–2 September 1992*, Washington DC, World Bank (Technical Paper, Biotechnology Series No. 1)

[24] Rissler and Mellon, op. cit.

[25] ibid.

[26] ODI, 1988, *Agricultural Biotechnology and the Third World*, London, Overseas Development Institute (Briefing Paper)

[27] M. Dibner, 1991, 'Tracking trends in US biotechnology', *Bio/Technology*, 9

[28] M. Greeley, 1992, *Agricultural Biotechnology, Poverty and Employment. The Policy Context and Research Priorities*, paper for the Technology and Employment Branch

of the International Labour Organization, Sussex (UK), Institute for Development Studies
[29] Herdt, op. cit.
[30] IRRI, 1996, op. cit.

9   Sustainable Agriculture

Agri cultura '. . . est scientia,
quae sint in quoque agro serenda ac facienda,
quo terra maximos perpetuo reddat fructus'
— Varro, *Rerum rusticarum* [1]

When he wrote these words, Marcus Terentius Varro, a Roman landowner of the first century BC, was eighty years old and had recently remarried. *Rerum rusticarum*, one of a number of Latin treatises on agriculture to survive to the present day, was written for his wife as a handbook of advice on how to run the estate he had purchased for her. In this passage he defines, for the first time, the concept of sustainability. Agriculture, he says, is 'a science, which teaches us what crops are to be planted in each kind of soil, and what operations are to be carried on, in order that the land may produce the highest yields in perpetuity'. As is common in the best of Roman writing, the definition is clear, elegant and succinct.

Unfortunately, Varro's original clarity of meaning has been lost. Sustainability has become a highly politicized term and, in the process, acquired a diversity of meanings [2]. Plant breeders, agronomists and other agriculturists interpret sustainability as the maintenance of the momentum of the Green Revolution. They equate it with achieving food sufficiency, and sustainable agriculture can embrace any means to that end. But for environmentalists the means are crucial: sustainable agriculture is a way of providing sufficient food without degrading natural resources. For economists, it represents an efficient, long-term use of resources, and for sociologists and anthropologists it embodies an agriculture that preserves traditional values and institutions. It has become an all-embracing term. Almost anything that is perceived as 'good' from the writer's perspective can fall under the umbrella of sustainable agriculture: organic farming, the small family farm, indigenous technical knowledge, biodiversity, integrated pest management, self-sufficiency, recycling and so on.

This diversity of interpretation is to be welcomed as part of a process of gaining consensus for radical change. Popular interest in sustainability was aroused by the report, published in 1987, of the World Commission on Environment and Development, chaired by the former Prime Minister of Norway, Mrs Gro Harlem Brundtland. The report drew the attention of politicians and the public, in both the developed and the developing world, to the threats to the world's survival posed by the way we treat our environment. The Brundtland Report's often-quoted definition of sustainable development – 'development that meets the need of the present without compromising the ability of future generations to meet their own needs' – was thus to be welcomed as a spur for political action [3]. But the subsequent debate has been confusing, as different interest groups have wrestled with the practical implications.

Much of the report was concerned with industrial and urban development although, as I have argued earlier in this book, agriculture is as much to blame for environmental degradation and is victim as well as culprit. The Brundtland definition in the context of agriculture is valuable as a policy statement but it is too abstract for the farmers, research scientists and extension workers who are trying to design new agricultural systems and develop new agricultural practices. For them a definition is needed which is scientific, is open to hypothesis-testing and experimentation, and is practicable.

Here, in theory and practice, lies the contribution of the second great revolution in modern biology, the emergence of ecology as a sophisticated discipline. The origins of ecology lie in the last century and are allied to the development of evolutionary theory. Indeed, Charles Darwin can be considered one of the first ecologists. In a passage in his *Journal* on the countries visited by HMS *Beagle* he describes the rich communities supported by the giant kelp beds off Tierra del Fuego. The leaves of the seaweed are covered with corallines, numerous shells and crustacea. 'On shaking the great entangled roots, a pile of small fish, cuttle fish, crabs of all orders, sea eggs, starfish, beautiful Holuthuriae, Planariae, and crawling nereidous animals of multitude of forms, all fall out together . . . Amidst the leaves of this plant numerous species of fish live, which nowhere else could find food or shelter; with their destruction the many cormorants and other fishing birds, the otters, seals, and porpoises, would soon perish also; and lastly, the Fuegan . . . perhaps cease to exist' [4]. In many ways the kelp community was like a tropical rainforest, the rich diversity of life and the complex web of food-chains supporting a very high level of productivity.

Until the middle of this century, ecology remained largely descriptive, often little more than natural history. But in recent years, the development of powerful hypotheses, the influence of mathematical tools and the design of appropriate laboratory and field experiments have transformed the natural history into a science [5]. Like most sciences it has its pure and applied sides, and its applications are now influencing many aspects of natural-resource management, including agriculture [6].

Population, community and ecosystem ecology have begun to provide a better understanding of the complex dynamics that arise within agriculture, for example in crop populations, in multiple-cropping systems and agroforestry and in range management. Some of the most fruitful work has come from collaboration between ecologists and agricultural scientists, often at the newer, so-called 'eco-regional' International Agricultural Research Centres, such as at the Centro Internacional de Agricultura Tropical (CIAT) in Colombia, but also at independent centres such as the Multiple Cropping Centre at Chiang Mai University where I and a group of Thai agronomists and agricultural scientists worked in the 1980s. We were attempting to understand, and improve upon, the complex agricultural systems that farmers had developed in the valleys of northern Thailand [7]. The concepts and definitions that follow are based on this work, which in turn was influenced by earlier ecological studies of the savanna ecosystems of Zimbabwe [8].

It takes little effort to recognize agricultural systems, such as the ricefields of northern Thailand, as modified ecological systems. Each field is formed, from the natural environment, by building up a ridge of earth that defines its boundary (Figure 9.1). Inside, the great diversity of the original wildlife is reduced to a limited set of crops, pests and weeds – although still retaining some of the natural elements, such as fish and predatory birds. The basic ecological processes remain the same:

− Competition between the rice and the weeds;
− Herbivory of the rice by the pests; and
− Predation of the pests by their natural enemies (and of the fish by the predatory birds).

165

But these ecological processes are now overlain and regulated by the agricultural processes of cultivation, subsidy (with fertilizers), control (of water, pests and diseases), and harvesting [9].

However, this is only a partial picture of the transformation. The agricultural processes are, in turn, regulated by economic and social decisions. Rice farmers cooperate or compete with one another and market, exchange or consume their produce. The resulting system is as much a socio-economic system as it is an ecological system, and has a socio-economic boundary although it is not as easy to define as the biophysical one of the earthen ridge. This new, complex, agro- socio- economic-ecological system, bounded in several dimensions, I call an agroecosystem.

More formally, an agroecosystem is 'an ecological and socio-economic system, comprising domesticated plants and/or animals and the people who husband them, intended for the purpose of producing food, fibre or other agricultural products' [10].

Agroecosystems defined in this way fall into a hierarchy. At the lowest level is the individual plant or animal and its immediate microenvironment (Figure 9.2). The next level is the crop or the herd or flock of animals,

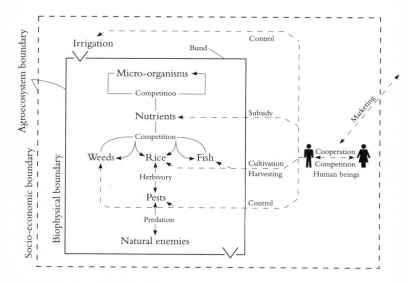

Figure 9.1 The ricefield as an agroecosystem [11]

contained within a field. The fields combine to form a farming system which is managed by a farm household. And the hierarchy continues upwards in a similar fashion.

Agroecosystems also have distinctive properties which can be measured and used as indicators of performance. For example, the productivity of a ricefield can be assessed – in terms of tons of rice per hectare. The stability of production can also be measured – how does production vary from year to year? So can its sustainability – how durable is the productivity? Varro's definition of the term sustainability, his use of the word *perpetuo*, equates it to persistence. The implicit question we ask is: will the agricultural system last, will it be productive not only in the immediate future, but over the longer term? Persistence, however, can only be measured in retrospect. Durability is a more practical concept, particularly if it can be assessed in terms of the forces that are likely to cause an agricultural system to collapse.

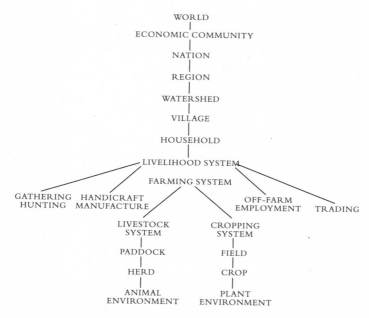

Figure 9.2 The hierarchy of agroecosystems [12]

This approach accords with common usage: sustainability in every day language refers to a capacity to maintain some activity in the face of a stress or shock — for example, to sustain a physical exercise, such as jogging or doing press-ups. On this analogy, agricultural sustainability is the ability of an agroecosystem to maintain productivity in the face of stress or shock. A stress may result from salinity, or pest attack, or erosion or debt — producing a frequent, sometimes continuous, adverse effect on productivity. A major event such as a disease outbreak or a drought or a sudden massive increase in the price of fertilizers would constitute a shock. As a response to the stress or shock, the productivity of the agroecosystem may not be affected, or it may fall and then eventually recover, or the system may collapse altogether (Figure 9.3). The advantage of this definition is that sustainability can be measured, or at least assessed. We also can begin to identify the components of sustainability — the intrinsic structure and processes in the agroecosystem, the nature and strengths of the likely stresses and shocks, and the human inputs which may be introduced to counter them.

One way of improving sustainability is to protect the agroecosystem from stress or shock. Nomadic pastoralists move their cattle to escape an impending drought. Farmers build a ridge to prevent flooding of crop-fields. In analogous fashion, a tariff wall, as has been erected by the European Community, may protect crop production from falling world prices. The alternative to protection is to take active countermeasures (Box 9.1).

A common stress experienced by a crop of rice, and indeed by most crops, is attack by pests or pathogens, or competition from weeds (see Chapter 11). The question we ask is: how resilient is the crop? Do the plants contain inherited resistance to attack, are natural enemies of the pests present, is the crop cultivated in such a way as to destroy the weeds? Most important, is the crop protected, by a broad-based genetic resistance, from attack by unexpected pests or new strains of pathogens? Spraying pesticides may be effective in controlling pests in the short run but often makes the long-term situation worse. More sustainable approaches rely on biological and integrated methods of control.

The stress that is experienced by all crops (and livestock populations) is the process of harvesting. The desirable product — grain, fruits, leaves, milk, wool, eggs — is removed and with it the nutrients. If the yield is to be sustainable, the nutrients have to be replaced (see Chapter 12). Usually

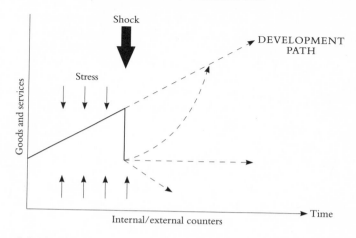

Figure 9.3 The dynamics of agricultural sustainability

we apply fertilizers to crop-fields, but sustainability can be enhanced by boosting natural fertility – growing nitrogen-fixing legumes or returning nutrients through composts. A common method of cultivation in the developing countries, particularly in the tropical uplands, is shifting cultivation (otherwise known as swidden cultivation or slash-and-burn). Forest is cleared and burned and a crop grown for several successive years until yields fall too far. The land is then allowed to revert to forest and the farmer opens up (shifts to) another piece of land. Once the natural fertility has recovered, the regrowth is cut and the cycle repeats itself. Sustainability here is crucially dependent on the respective lengths of the cropping and the fallow periods. In the example in Figure 9.4, when up to eight crops are grown in the cropping phase, the system goes back to mature forest but beyond eight crops it degrades to an unproductive grassland. Grazing systems can behave in a similar fashion [13]. Increasing the number of livestock on a range raises productivity but also stresses the vegetation. At a certain intensity of stress, the vegetation collapses and the grazing system moves to a new level of productivity, much lower than before.

Analogous countermeasures to the effect of stress and shock can be taken at higher levels. The response of a farmer to growing debt may be

Box 9.1 Agricultural technologies that have high potential sustainability [14]

Intercropping	The growing of two or more crops simultaneously on the same piece of land. Benefits arise because crops exploit different resources, or mutually interact with one another. If one crop is a legume it may provide nutrients for the other. The inter-actions may also serve to control pests and weeds.
Rotations	The growing of two or more crops in sequence on the same piece of land. Benefits are similar to those arising from inter-cropping.
Agroforestry	A form of intercropping in which annual herbaceous crops are grown interspersed with perennial trees or shrubs. The deeper-rooted trees can often exploit water and nutrients not available to the herbs. The trees may also provide shade and mulch, while the ground cover of herbs reduces weeds and prevents erosion.
Sylvo-pasture	Similar to agroforestry, but combining trees with grassland and other fodder species on which livestock graze. The mixture of browse, grass and herbs often supports mixed livestock.
Green manuring	The growing of legumes and other plants in order to fix nitrogen and then incorporating them in the soil for the following crop. Commonly used green manures are *Sesbania* and the fern *Azolla*, which contains nitrogen-fixing, blue-green algae.
Conservation tillage	Systems of minimum tillage or no-tillage, in which the seed is placed directly in the soil with little or no preparatory cultivation. This reduces the amount of soil disturbance and so lessens run-off and loss of sediments and nutrients.
Biological control	The use of natural enemies, parasites or predators, to control pests. If the pest is exotic these enemies may be imported from the country of origin of the pest; if indigenous, various techniques are used to augment the numbers of the existing natural enemies.
Integrated pest management	The use of all appropriate techniques of controlling pests in an integrated manner that enhances rather than destroys natural controls. If pesticides are part of the programme, they are used sparingly and selectively, so as not to interfere with natural enemies.

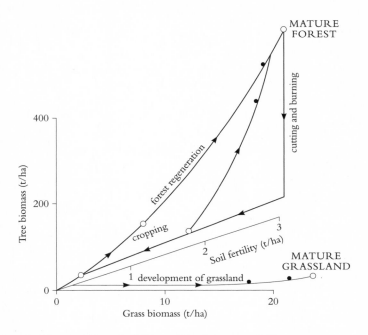

Figure 9.4 Sustainability of a shifting cultivation system [15]

to switch to a less risky crop/livestock combination or to one requiring lower inputs. A village stressed by the loss of young people emigrating in search of more lucrative work may respond by adopting more labour-saving techniques of cultivation. Similarly, a district may counter rising transport costs by a switch to higher-value, lower-volume products; a region may react to a widespread drought by establishing a network of famine-relief stores as a protection against future droughts; and a nation may respond to increasing competition by changing its export-crop priorities so as to exploit its comparative advantage with respect to other nations.

A critical issue is how much should sustainability rely on outside inputs – such as fertilizers and pesticides, in contrast to the internal resources that are available on the farm or within the community and immediate environment [16]. Internal resources are typically natural resources. They include:

- The natural parasites and predators of pests;
- Algae, bacteria and green manures supplying nitrogen;
- Agroforestry and cropping systems reducing erosion;
- Underexploited wild tree and fish species;
- Indigenous crop varieties with tolerance to salts and toxins.

These resources are inherently renewable and thus have the potential to be used on a sustained basis, often 'free of charge'. By contrast, external resources, such as irrigation water from a distant source, or synthetic fertilizers and chemical pesticides, are not locally renewable and have to be obtained from outside the farm or community. They normally have to be purchased, and at prices outside the control of the farmer. Even where there is no direct monetary cost, there may be costs in terms of the time allocated for obtaining resources – purchasing seeds and other inputs from local distribution centres, or searching for fuel-wood and fodder from distant areas, or carrying water from the nearest well.

Dependence on external resources is not only often costly and risky – because of sudden changes in price and availability – it may also lead to fundamental changes in the farming system that make it more vulnerable to the vagaries of the local environment. This is one explanation of the failure to adopt the Green Revolution 'packages' of hybrid seeds, fertilizers and pesticides in lower-potential environments. Such packages are often less suited to these environments compared to the lower-yielding, yet better-adapted, internal resources used in traditional farming systems.

Labour, capital, machinery and management can be either internal or external to the farm. When these resources are primarily internal – for example a family-owned and operated farm – then households have a greater degree of control over decisions concerning the allocation of resources and their long-term management. In Nepal production is highest on land which is cultivated by farmers who own the land and lowest on lands tilled by informal tenants on a contract basis. Where a farmer both owns and rents land, production is higher on the former. More importantly, landowners who cultivate their own land have a greater incentive to manage it sustainably. In contrast, tenants are unlikely, on their own, to take an interest in the long-term productivity of the land they are working.

However, the contrast between internal and external resources is not as stark as I have portrayed it. Internal resources are usually preferable, if they are available and effective, but often external resources are necessary, or even superior. Much depends on the nature of the local circumstances and, in particular, the actual and potential threats to sustainability. In the last chapter I described the likely benefits of genetic engineering. These clearly extend to the improvement of sustainability. New crop varieties engineered for pest or pathogen resistance or with the capacity to fix their own nitrogen could dramatically transform agricultural systems that might otherwise decline or collapse. Genetically engineered plants are an external resource, and can be costly if farmers have to purchase new seed each year. But the costs and risk can be reduced if the desirable genes are engineered into traditional varieties and made freely available.

The choice between external and internal resources is, of course, not only determined by considerations of sustainability. As I indicated in the opening paragraphs, sustainability is but one property, or indicator of the performance, of an agroecosystem. Altogether there are four principal indicators (Figure 9.5):

Productivity: The output of valued product per unit of resource input.

Stability: The constancy of productivity in the face of the normal fluctuations and cycles in the surrounding environment. (Usually measured from a time series by the coefficient of variation in productivity.)

Sustainability: The ability of the agroecosystem to maintain productivity when subject to a stress or shock;

Equitability: The evenness of distribution of the productivity of the agroecosystem among the human beneficiaries, i.e. the level of equity that is generated. (A common measure is a Lorenz curve.)

In this book, I have used various measures of productivity, reflecting different combinations of output and input. Most agronomists refer to yield, expressed as kilograms of grain, tubers or meat per hectare, or per kilogram of nitrogen or per hour of human effort. Nutritionists may be more interested in the output of calories, proteins or vitamins, while economists will measure the monetary value of the agricultural production at the market. In the last case productivity is expressed as income less

PRODUCTIVITY

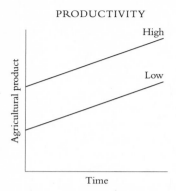

Productivity: The output of valued
product per unit of resource input.

STABILITY

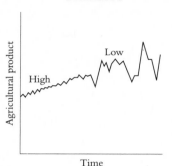

Stability: The constancy of productivity
in the surrounding enviroment (usually
measured from a time series by the
coefficient of variation in productivity).

SUSTAINABILITY

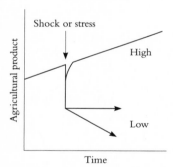

Sustainability: The ability of the
agroecosystem to maintain productivity
when subject to a stress or shock.

EQUITABILITY

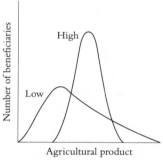

Equitability: The eveness of distribution
of the agroecosystem among the human
beneficiaries, i.e. the level of equity that
is generated (a common measure is a
Lorentz curve).

Figure 9.5 Indicators of performance [17]

expenditure, that is, as net income or profit. We also frequently measure
total production of agricultural goods and services per household or per
region or nation.

Over time productivity may exhibit a dominant trend – rising, falling or remaining constant. It will also fluctuate about the trend. The yield of a crop is likely to mirror the variability in the climate. Income may also fluctuate, not only reflecting changes in yield but also variations in the market price of inputs, such as labour, fertilizers and pesticides, and of the product. Stability is a measure of such variability in the face of these relatively minor and commonplace disturbing forces. By contrast, sustainability is concerned with forces that are insidious or are rarer, and less expected, so that agroecosystems are likely to have fewer or less-well-developed defences.

Productivity, stability and sustainability adequately measure how much an agroecosystem produces and is likely to produce over time. An African village that has a high, stable yield of sorghum, using practices and varieties that are broadly resistant to pests and diseases, might be regarded as more successful than another village having lower, less stable and sustainable yields. However, it is not only the pattern of production that is important, but also the pattern of consumption. Who benefits from the high, stable and sustainable production? How is the harvested sorghum, or the income from the sorghum, distributed among the people of the village? Is it evenly shared or do some villagers benefit more than others? Equitability describes this pattern of distribution of productivity. Commonly, equitability describes the distribution of the overall production, that is the total goods and services produced by an agroecosystem. In subsistence farming the producers are all consumers, but the higher the agroecosystem in the hierarchy and the greater the degree of commercialization, the more do non-producers benefit (see Chapter 5).

Inevitably there are trade-offs between the levels of these indicators. For example, a large-scale irrigation project may achieve greater overall productivity, yet be at the expense of sustainability and equitability. Similarly, too much emphasis on equitability may inhibit productivity. In many respects the Green Revolution as a process has favoured high productivity at the expense of the other indicators (Box 9.2).

At different periods in history, and in differing circumstances, one or more properties have been elevated as targets of development. In the Green Revolution this was a conscious choice. As Randolph Barker, Robert Herdt and Beth Rose put it: 'If the modest resources available to the international agricultural research system in the 1960s had been concentrated in the less favorable environments, it is likely that no major

Box 9.2 Effect of the Green Revolution on agroecosystem indicators

Productivity	Higher cereal yields and production in most regions
	Higher per capita cereal yields and calorie supply in most regions
	– but lower in Sub-Saharan Africa
	Higher food production, especially livestock production, in some regions
	Higher farm incomes in Green Revolution lands
	– but not in most lower potential lands
	More employment and higher real wages
	– but not where mechanization coincides with growing labour supply
Stability	Greater variance in yields and production in some regions
	– due to pest and pathogen attack
	– or global warming-induced, climate variability
Sustainability	Greater pest and disease resistance
	– but severe new outbreaks on some crops
	– and increased morbidity and mortality from pesticides
	Greater reliance on inorganic nitrogen fertilizers with risk of restrictions
	Risk of damage and disruption due to acid rain, ozone pollution, penetration of UV light, and global warming
	Loss of soil structure and micronutrients
	Increased soil toxicity, waterlogging and salinity
Equitability	Benefits disproportionately to landowners and providers of inputs
	Declining real wages, increased unemployment and increased landlessness in some regions
	Persistence of high levels of chronic undernutrition and malnutrition

breakthrough would have been made' [18]. However, as I argued in Chapter 3, we are entering a new phase of agricultural development, where much greater attention will have to be paid to increasing productivity while, at the same time, raising stability, sustainability and equity. The object of the Doubly Green Revolution is to seek to explicitly minimize the trade-offs.

How this can be achieved I shall explore in subsequent chapters, but as an illustration of the principles and processes involved it is worth looking, as Otto Soemarwoto of Padjadjaran University in Indonesia and I have done, at a traditional agroecosystem – the home garden – in which, through centuries of manipulation, the trade-offs have been effectively minimized.

Home gardens are one of the oldest forms of farming system and may have been the first agricultural system to emerge in hunting and gathering societies [19]. Today, home or kitchen gardens are particularly well developed on the island of Java in Indonesia, where they are called *pekarangan* [20]. Their immediately notable characteristic is their great diversity relative to their size: they usually take up little more than half a hectare around the farmer's house. In one Javanese home garden fifty-six different species of useful plants were found, some used for food, others as condiments and spices, some for medicine and others as feed for the livestock – a cow and a goat, some chickens or ducks, and fish in the garden pond. Much is for household consumption, but some is bartered with neighbours and some is sold. The plants are grown in intricate relationships with one another: close to the ground are vegetables, sweet potatoes, taro and spices; in the next layer are bananas, papayas and other fruits; a couple of metres above are soursop, guava and cloves, while emerging through the canopy are coconuts and timber trees, such as *Albizzia*. So dense is the planting that to a casual observer the garden seems like a miniature forest. The diversity is in contrast to the adjacent, much-simplified ricefield systems where the only crop is rice, perhaps with some edible weeds and fish. Closer analysis shows the high diversity in the home garden is matched by high levels of productivity, stability, sustainability and equitability. A comparable ricefield has higher gross income but otherwise its productivity is not as high and its other indicators are considerably lower (Box 9.3).

Part of the reason for the minimal trade-off in the home garden is the inherent diversity. It helps stabilize production, buffers against stress and shock and contributes to a more valued level of production. But equally important is the intimate nature of the home garden. The close attention that is possible from family labour ensures a high degree of stability and sustainability and the link between the garden and the traditional culture leads to an equitable distribution of the diverse products.

The point of this illustration is not, of course, to suggest that ricefields

Box 9.3 **System properties of the home garden when compared with a ricefield** [21]

	Home garden	*Ricefield*
Productivity	Higher standing biomass Higher net income (lower inputs) Greater variety of production (food, medicines, fuel-wood)	Higher gross income
Stability	Year-round production ('living granary') Higher year-to-year stability	Seasonal production Vulnerable to climatic and disease variation
Sustainability	Maintenance of soil fertility Protection from soil erosion	Heavy pest and disease attack
Equitability	Home gardens in most households Barter of products	Product to landowners

should be converted to home gardens. Rather it is, as I said at the outset, to underline certain principles and processes. The first is the importance of maintaining and, wherever possible, enhancing diversity. More diverse agroecosystems tend to be more sustainable and, often, more productive than systems which are otherwise comparable. Second is the crucial role played by the farm household, in making decisions and choices. Successful agriculture depends as much on the farm household as it does on crops and livestock and the ecological processes that link them.

Farm households are usually complex entities, consisting of different generations, men and women and their offspring, and often extending through brothers and sisters and their families to embrace a considerable number of people. While agriculture has sometimes been seen as essentially a male activity, and it is still common for agricultural extension workers to be male and to only interact with men, in recent years the interest among development workers in gender issues has provided a growing understanding of the critical importance of women in farm households.

Agnes Quisumbing and her colleagues at IFPRI distinguish three contributions which women make to food security: food production, access to food, and nutrition security [22]. Women account for half the production of food in developing countries. In Sub-Saharan Africa, where women and men customarily farm separate plots, it is as high as 75 per cent, the men concentrating their efforts on cash crops [23]. African women are also responsible for 90 per cent of the work involved in processing food and collecting fuel-wood and water. Men play a greater role in food production in Asia and Latin America, but women are still major contributors. Women are mostly responsible for transplanting, weeding and harvesting in rice production, and they are normally the cultivators of home gardens and vegetable plots.

As I indicated in Chapter 5, while the higher production and cropping intensity of the early Green Revolution increased the demand for women's labour, women have tended to subsequently be displaced by mechanization, particularly in harvesting and milling, with severe effects on poor households. Lack of employment and incomes for women not only result in lower overall household income, they crucially affect children's access to food. Women typically spend a high proportion of their income on food and health care for children. A study carried out in Rwanda by Joachim von Braun and colleagues at IFPRI showed a close relationship between women's income and calorie consumption in the household, even though women earned considerably less than men [24]. Significantly, there were no severely malnourished children in female-headed households.

Nutrition of children is also closely related to health care. Children are better nourished where breast feeding is practised and attention is paid to providing nutritious food for weaned infants and to the maintenance of hygiene. This is predominantly, and usually exclusively, women's work, yet if they are active as farm workers or earning off-farm income there are severe constraints on their time. A number of studies in Africa and Asia have shown women to be spending less than an hour a day in direct child care [25]. As Jeffery Leonard of IFPRI points out:

. . . women's multiple roles in poor households perpetually conflict with each other. For example, increased time spent on out-of-household chores or non-household employment can directly reduce the time women have for child-rearing and other household duties. Conversely, the time and energy devoted to just gathering fuel and water increasingly has been recognised as a major impediment to efforts to increase women's contributions to food production, household income, or family welfare [26].

In Nepal, where severe deforestation has forced women to go further afield and spend more time in collecting fuel-wood, women have less time for other tasks and, as a direct result, child nutrition has worsened [27].

It is easy from a developed-country, urban perspective to see a livelihood, the means of securing a living, as a relatively simple affair – a single head of household bringing in a regular, assured income (although this is changing, with families increasingly relying on two incomes and more flexible and varied patterns of work). But for the rural poor in the developing countries livelihoods are always highly complicated affairs, dependent on an intricate mix of assets and resources, rights and obligations, and the skills of the farm household (Chapter 15) [5]. A livelihood may be constructed from:

– Land on which crops or livestock are husbanded; or
– Natural resources – timber, fuel-wood, wild plants, fish and other wild animals – which may be harvested; or
– Opportunities for off-farm employment; or
– Skills employed on the farm in manufacture of handicrafts; or, most commonly,
– Some combination of these (with surprisingly few exceptions, developing-country farmers do not rely exclusively on farming).

In theory, households decide on livelihood goals – the balance between higher productivity, greater sustainability or improved equity – and then determine what is the optimal mix of activities. But this is not a straightforward process. It depends on the skills and resources at their disposal, and on their environmental and social circumstances. Since they are integral components of communities, they are bound by traditional customs and systems of rights and obligations. Yet their decisions are also determined by their perceptions of the present and future world in which they live, and by the opportunities which appear to be offered by new technologies.

Few comparative studies of livelihoods have been carried out and none, that I can find, of farmers in the Green Revolution lands. However, a livelihood analysis which illuminates the nature of the choices confronting poor rural households has been conducted for four Amerindian groups in Central Brazil [28]. The groups practise a combination of gardening, hunting and fishing and Figure 9.6 shows the yields per person-hour for these different productive enterprises, together with the time spent on each.

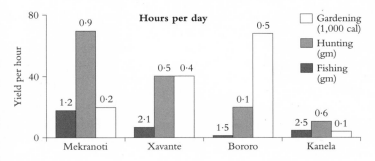

Figure 9.6 Productivities of four Brazilian Amerindian livelihoods [29]

In the case of fishing the relationship between yield and effort is fairly straightforward: the higher the productivity the greater the time spent fishing. For gardening and hunting the relationship is more complex. The gardens of the Mekranoti, for example, produce far more calories than they could possibly eat – over 21,000 calories per person per day; the excess is stored as insurance against bad years, or kept to feed visitors from outside, and this appears to be why they spend less time gardening than the others. Hunting is also more productive for the Mekranoti, but here they devote more time to hunting than the other groups, possibly because they have the time to do so and appreciate the quality of a high-protein diet.

The Kanela have a high population density and live in a much poorer habitat. Hunting and fishing are very unproductive and they spend a relatively large amount of time gardening, concentrating on protein-poor manioc to provide the calories they require. They also spend more time than the other groups in producing handicrafts for sale or working for wages, in order to make up the protein shortfall and to satisfy other requirements. The Xavanate lie between these two extremes.

The Bororo also live in a poor environment for gardening, only producing 2,000 calories per person per day, but fish production is high. They sell some fish for high-calorie foods, but the market is far away and they get a low price. Fishing is also very hard work and they suffer from a high rate of illness and invalidism; otherwise they might put even more time into fishing.

Such an analysis may seem of purely academic interest, but it clearly reveals some of the practical constraints to development. The sustainability ᴏf these livelihoods depends on their diversity, and potential innovations – such as new crops to improve the vegetable protein intake of the Kanela, or the calorie intake of the Bororo – must not conflict, in terms of labour demand, with the existing patterns of profitable activity. In my view, livelihood analyses of this kind are essential first steps in any development programme, and it is a matter of concern that so few have been undertaken.

The approach outlined in this chapter, with its emphasis on ecological processes and the complexities of household decision-making, may seem very distant from the molecular technology underlying genetic engineering which I described in the previous chapter; nevertheless not only are they both revolutionary in their potential impact, they are interconnected. They are having considerable impact on the conduct and techniques of laboratory and field research, and, more important, are providing novel ways of thinking and enquiring about biological, agricultural and socio-economic phenomena, bringing new system perspectives and enhancing our capacity to define critical answerable questions.

They are both crucial to the success of the Doubly Green Revolution. They are not alternatives. Indeed they are complementary, providing the means whereby farmers, field and laboratory scientists can collaborate in identifying and answering the research questions posed by the socio-economic needs of poor households. The way forward lies in harnessing the power of modern technology, but harnessing it wisely in the interests of the poor and hungry and with respect for the environment in which we live. We need a shared vision based, above all, on partnership, among scientists and between scientists and the rural poor.

Notes

[1] W. D. Hooper and H. B. Ash, 1935, *Marcus Porcius Cato on Agriculture. Marcus Terentius Varro on Agriculture*, Cambridge, Mass., Harvard University Press/London, Heinemann (Loeb Classical Library). (The English version in the first paragraph is my translation.)

[2] G. R. Conway and E. B. Barbier, 1990, *After the Green Revolution: Sustainable Agriculture for Development*, London, Earthscan

[3] World Commission on Environment and Development, 1987, *Our Common Future*, Oxford, Oxford University Press

[4] C. Darwin, 1890, *A Naturalist's Voyage: Journal of Researches into the Natural*

History and Geology of the Countries Visited During the Voyage of H.M.S. 'Beagle' round the World, under the Command of Capt. Fitz Roy, R.N. (new edn), London, John Murray

[5] R. M. May (ed.), 1981, *Theoretical Ecology: Principles and Applications* (2nd edn), Oxford, Blackwell Scientific; M. Begon, J. L. Harper and C. R. Townsend, 1990, *Ecology: Individuals, Populations and Communities* (2nd edn), Oxford, Blackwell Scientific

[6] M. A. Altieri, 1995, *Agroecology: The Science of Sustainable Agriculture* (2nd edn), Boulder, Colo., Westview Press/London, Intermediate Technology; R. Lowrance, B. R. Stinner and G. J. House (eds.), 1984, *Agricultural Ecosystems: Unifying Concepts*, Chichester (UK), Wiley & Sons; National Research Council, 1989, *Alternative Agriculture*, Committee on the Role of Alternative Farming Methods in Modern Agriculture, Washington DC, National Academy Press; 1993, *Sustainable Agriculture and the Environment in the Humid Tropics*, Committee on Sustainable Agriculture and the Environment in the Humid Tropics, Washington DC, National Academy Press; M. G. Paoletti, B. R. Stinner and G. G. Lorenzoni (eds.), 1989, *Agricultural Ecology and Environment. Proceedings of an International Symposium on Agricultural Ecology and Environment, Padova, Italy, 5–7 April 1988*, Amsterdam, Elsevier; J. Tivy, 1990, *Agricultural Ecology*, New York, NY, Longman Scientific and Technical

[7] P. Gypmantasiri, A. Wiboonpongse, B. Rerkasem, I. Craig, K. Rerkasem, L. Ganjanapan, M. Titayawan, M. Seetisarn, P. Thani, R. Jaisaard, S. Ongprasert, T. Radnachaless and G. R. Conway, 1980, *An Interdisciplinary Perspective of Cropping Systems in the Chiang Mai Valley: Key Questions for Research*, Chiang Mai (Thailand), Faculty of Agriculture, University of Chiang Mai

[8] B. H. Walker, G. A. Norton, N. D. Barlow, G. R. Conway, M. Birley and H. N. Comins, 1978, 'A procedure for multidisciplinary ecosystem research with reference to the South African Savanna Ecosystem Project', *Journal of Applied Ecology*, 15, 408–502

[9] Lowrance *et al.*, op. cit.; C. R. W. Spedding, 1975, *The Biology of Agricultural Systems*, London, Academic Press

[10] G. R. Conway, 1987, 'The properties of agroecosystems', *Agricultural Systems*, 24, 95–117

[11] ibid.

[12] ibid.

[13] I. Noy-Meir, 1975, 'Stability of grazing systems: an application of predator–prey graphs', *Journal of Ecology*, 63, 459–81

[14] Conway and Barbier, op. cit.

[15] B. R. Trenbath, G. R. Conway and I. A. Craig, 1990, 'Threats to sustainability in intensified agricultural systems: analysis and implications for management', in S. R. Gliessman (ed.), *Agroecology: Researching the Ecological Basis of Sustainable Agriculture*, New York, NY, Springer-Verlag, pp. 337–66

[16] C. A. Francis and J. A. King, 1988, 'Cropping systems based on farm-derived, renewable resources', *Agricultural Systems*, 27, 67–75

[17] G. R. Conway, 1987, *Helping Poor Farmers – A Review of Foundation Activities in Farming Systems and Agroecosystems Research and Development*, New York, NY, Ford Foundation

[18] R. Barker, R. W. Herdt and B. Rose, 1985, *The Rice Economy of Asia*, Washington DC, Resources for the Future

[19] I. D. Hoogerbrugge and L. Fresco, 1993, *Homegarden Systems: Agricultural Characteristics and Challenges*, Sustainable Agriculture Programme, London, International Institute for Environment and Development

[20] O. Soemarwoto and G. R. Conway, 1991, 'The Javanese homegarden', *Journal for Farming Systems Research and Extension*, 2, 95–118; Altieri, op. cit.

[21] Soemarwoto and Conway, op. cit.

[22] A. R. Quisumbing, L. R. Brown, H. S. Feldstein, L. Haddad and C. Pen-a, 1995, *Women: The Key to Food Security*, Washington, DC, International Food Policy Research Institute (Food Policy Report)

[23] FAO, 1985, *Women and Developing Agriculture*, Rome (Italy), Food and Agriculture Organization (Women in Agriculture Series No. 4)

[24] J. von Braun, H. de Haen and J. Blanken, 1991, *Commercialization of Agriculture under Population Pressure: Effects on Production, Consumption and Nutrition in Rwanda*, Washington DC, International Food Policy Research Institute (Research Report 85)

[25] L. R. Brown and L. Hadad, 1994, *Time Allocation Patterns and Time Burdens: A Gendered Analysis of Seven Countries*, Washington DC, International Food Policy Research Institute

[26] H. J. Leonard, 1989, 'Overview: environment and the poor', in H. J. Leonard, *Environment and the Poor: Development Strategies for a Common Agenda*, Overseas Development Council (U.S.–Third World Policy Perspectives No. 11), referring to N. Sadik, 1988, 'Women as resource managers', in UNPF, *State of the World Population, 1985*, New York, NY, United Nations Population Fund

[27] S. K. Kumar and D. Hotchkiss, 1988, *Consequences of Deforestation for Women's Time Allocation, Agricultural Production, and Nutrition in Hill Areas of Nepal*, Washington DC, International Food Policy Research Institute (IFPRI Research Report 69)

[28] D. Werner, N. M. Flowers, M. L. Ritter and D. R. Grass, 1979, 'Subsistence productivity and hunting effort in native South America', *Human Ecology*, 7, 303–15

[29] Werner *et al.*, op. cit.

10   Partnerships

> The question is – how often, and in what circumstances, with what
> learning and understanding, are the poor man, and more, the poor
> woman and child, met, listened to, and understood; and how often
> are the distant effects of decisions and actions, and of indecisions and
> non-actions, reflected upon.
>   – Robert Chambers, N. C. Saxena and Tushaar Shah,
>   *To the Hands of the Poor* [1]

Farmers have been experimenters since the beginning of agriculture [2].
Hunters and gatherers had long learned to use fire as a means of stimulating
the growth of tubers and other food plants, and of grass to attract game.
Plant selection began when people found they could encourage favoured
fruiting trees by clearing their competitive neighbours, but the first steps
towards intensive plant breeding were taken when an individual, probably
a woman rather than a man, deliberately sowed a seed from a high-yielding
plant somewhere near the dwelling and observed it grow to maturity.
Besides producing new varieties, experimentation also resulted in whole
systems of agriculture: shifting cultivation, rice terracing, home gardens,
irrigated agriculture, the Mediterranean trio of wheat, olives and vines,
the Latin American multiple-cropping of maize, beans and squashes, and
various forms of integrated crop–livestock agriculture.

For most of history, farmers have been the primary innovators and
experimenters. As is evident from their writings, the Romans analysed the
structure and functions of agricultural systems in a scientific manner. They
also described the process of experimentation. Marcus Terentius Varro,
whom I quoted at the beginning of the last chapter, urged farmers to both
'imitate others and attempt by experiment to do some things in a different
way. Following not chance but some system: as, for instance, if we plough
a second time, more or less deeply than others, to see what effect this will
have' [3]. The great agricultural revolution of Britain in the late eighteenth

century was led by farmers. Jethro Tull is famous for his invention of the corn drill, Charles Townsend for the introduction of turnips, Thomas Coke for the Norfolk Four Course Rotation and Robert Bakewell for the selective breeding of livestock. However, as Jules Pretty points out, they were well-educated landowners who had read the Latin texts and understood the basic principles of sustainable agriculture and set about popularizing them; they were not the real innovators [4]. Over the previous hundred years, numerous 'unknown' farmers had been developing and propagating new techniques through an informal process of rural tours and surveys, farmers' groups and societies, open days, training and publications. The professionalization of agricultural research only properly began in the nineteenth century, notably following the creation of the Rothamsted Experimental Station in England in 1843 and of the Land Grant Colleges in the USA, although, even then, farmers were well represented on the boards of management, and research programmes were highly responsive to farmers' needs.

Research in the developing countries, under colonial rule, was, inevitably, of a top-down nature, with a strong emphasis on export crops. Little changed after independence. The top-down tendency was reinforced by the Green Revolution, despite the shift in emphasis to food crops. Although the early work in Mexico, in the 1940s and 1950s, on the breeding of new disease-resistant wheat varieties, was concerned with local adaptability, the realization by western scientists of the enormous potential of the dwarfing genes present in East Asian wheat and rice germ plasm led to a quest for varieties which would perform well, irrespective of local conditions. Inevitably, the richness of farmers' indigenous knowledge and their capacity to experiment was either ignored or downplayed [5].

Farmers usually hold strong, and often insightful, opinions on the qualities of crop types and livestock breeds. New varieties are often adversely compared with traditional varieties, with which they have long familiarity. In northern Thailand, highly commercial farmers grow as a first crop, for their own consumption, numerous varieties of the traditional sticky (non-glutinous) rice which has been part of their culture for hundreds of years, although later in the year they may plant new varieties for sale [6]. These are rarely simply prejudices, however; if asked farmers can identify attributes, compare positive and negative features and rank varieties placed in front of them.

Scientists at CIAT in Colombia, aware that the varieties they had developed according to research-station criteria were often not accepted, asked farmers to rank the grain from bush beans and explain their reasons [7]. The results produced rankings very different from those of the breeders themselves, and also revealed a difference in preferences between men and women, the latter choosing smaller, better-flavoured grains, while the men preferred the larger grains which command good prices in the market. A similar exercise which I carried out with Ethiopian farmers produced a ranking of tree species for cultivation. A set of six species was assessed by presenting them to the farmers in every pair-wise combination. The farmers were asked to indicate which they would prefer if they could only grow one of the pair and to give the reasons. Their accumulated ranking revealed the great range of usage for which different species were suitable [8]. When the process was repeated with a group of foresters, the ranking was very different. The foresters emphasized the ease and reliability of species in the nursery, which was their main responsibility, while the farmers placed a higher premium on versatility (Box 10.1).

Box 10.1 **Farmers' ranking of cultivated trees in Wollo, Ethiopia** [9]

1. African olive	Diverse utilization, including digging sticks, yoke and other plough parts, hoes, ox handles; incense from leaves; not attacked by termites; no smoke when used as firewood
2. *Eucalyptus camaldulensis*	Easy to split, straight, strong for construction, durable; easy to make charcoal
3. *E. globulis*	High elasticity; farming implements, good for holding nails; firewood but difficult to make charcoal
4. Juniper	Window and door timber; chair-making
5. White acacia	Housebuilding
6. Croton	Door construction; but smoky as firewood

Breeders usually try out their selections in farmers' fields to determine acceptability. But this has customarily occurred at the end of the breeding process when many of the key decisions have already been made. What has changed in recent years is a greater involvement of farmers earlier in the process, eliciting not only reactions but positive inputs into the determination of breeding goals. The next stage at CIAT was to encourage farmers to take seeds away and grow them in trials on their own land. At the end of the trials they produced overall rankings not only on grain quality but the performance of the plants (Box 10.2). Yield was not the dominant criterion. The farmers placed much greater emphasis on marketing, resistance to pests and disease and labour requirements.

In Rwanda a five-year experiment conducted by CIAT and by ISAR (Institut des Sciences Agronomiques de Rwanda) progressively involved farmers at even earlier stages in the breeding process [10]. Beans are a key component of the Rwandan diet, providing 65 per cent of the protein and 35 per cent of the calories, and are grown by virtually all farmers. There is an extraordinary range of local varieties – over 550 identified – and farmers (mostly women) are adept at developing local mixtures which breeders have difficulty in bettering. In the first phase of the experiment, teams of expert farmers were asked to evaluate some fifteen varieties, two to four seasons before normal on-farm testing. This revealed new criteria, for example the ability of varieties to perform well when grown under bananas, and also made breeders more aware of the range of expertise among the farmers. Some women were particularly astute at distinguishing among different criteria.

In the second phase, farmers were asked to assess a trial of about eighty lines, using their own criteria to reduce the number of lines. Farmers tagged favoured varieties on the station with coloured ribbons. After three years, twenty to twenty-five lines were taken to field trials. Two approaches to the trials were tried. In one, the research scientists drew up standard protocols (varieties sown in lines, at given densities) and the farmers were invited to assess the results. The researchers gained valuable feedback, but the process of adaptive testing and diffusion was slow. In the alternative model, the local communities determined the way the trials were conducted. A core group of farmers divided up the varieties and tested them on individual plots. The group then made selections and subsequently was responsible for multiplying and diffusing the most promising varieties. This experience led the plant breeders to conclude that 'the standard breeding models may not be using each partner's, breeder's and farmer's talents to best advantage, particularly in areas marked by marginal, heterogeneous environments. Breeders' unique expertise lies in their capacity to generate new scientific variability. Farmers do cross and select, but at an extremely

Box 10.2 **Farmers' preferences of bush beans in Colombia** [11]

Variety ranking	*Positive features*	*Negative features*
BAT-1297 By breeders: 10th, bottom of the list By farmers: 2nd	High-yielding, profitable, good flavour, resistance to pests and diseases, drought resistance, good germination, grain swells on cooking	Small grain, later variety
A–486 Breeders: 2nd Farmers: 6th	Nice grain colour and size, delicious, yields well, early	Quickly infested by storage pests, grain changes colour after harvest and is difficult to market
ANTIQUA BL–40 Breeders: 5th Farmers: least acceptable, 8th	Yields well	Variable grain colour makes marketing difficult, affected by disease, sprawling habit makes weeding difficult, many small and immature pods, some rotten at harvest, very late

slow rate: scientific breeding accelerates the process ... In turn, the finishing of the product, targeting the variety to a particular production system, can and should be left to farmers' (Box 10.3) [12]. There is now a considerable experience of participatory plant breeding programmes with, in some cases, farmers active in assessing breeding lines as early as the second- or third-generation crosses [13].

This experience and others has led the IARCs to pay more attention to farmers' capacity to experiment. It has long been the practice to place trial plots in farmers' fields, but often the farmers are simply used as labourers, being given little knowledge of the purpose of the trial. (I remember in West Africa visiting a trial of alley cropping – intercropping

THE DOUBLY GREEN REVOLUTION

Box 10.3 **Contributions to the breeding process** [14]

Breeders	*Farmers*
Create most new genetic variability	Create some new genetic variability
Make accessible wide range of germ plasm (local and exotic)	Target for agronomic conditions (performance)
Screen large amounts of material for minimum criteria	Target for socio-economic circumstances (preference)
Screen for key stresses invisible to farmers	

---

of maize and a legume tree. The farmer proudly pointed to a monoculture of maize but said it had been hard work because of the large number of small trees that kept emerging.) This disengaged approach has begun to change as more confidence has been gained in participatory research. Jacqueline Ashby and her colleagues at CIAT have established 'innovators' workshops' in which farmers design and evaluate experiments. One trial tackled the problem of lack of stakes for climbing snap beans. The farmers suggested growing the beans after tomatoes, so exploiting the tomato stakes and the residual fertilizer, and then chose two snap bean varieties as appropriate for the new system [15].

David Millar, who works for the Tamale Archdiocesan Agricultural Programme, asserts that there is no farmer in northern Ghana who is not in some way experimenting [16]. Some are pursuing curiosity experiments. One farmer, Dachil, had travelled to southern Ghana and brought back cocoyams which naturally grow in the forest. He planted them in his yard under the shade of a mango tree. 'If the results are good, my next step will be to set up a small garden on my farm . . . I am just curious to find out everything I can about the crop.' Other farmers are trying to solve problems. Millar describes the trials designed by Nafa and his brothers using different forms of crop rotation to eliminate a notorious weed, *Striga*. And large numbers of farmers in the region are engaged in adapting introduced technologies.

A classic example of technology adaptation was the worldwide response to the introduction of new technologies for potato storage developed by the International Potato Centre (CIP) [17]. The technologies were based on observations of the success of some farmers who, contrary to the normal practice of storing potatoes in the dark, used diffused light. CIP produced a package of recommendations which was then introduced to some twenty-five countries. But adoption did not proceed as expected. Virtually all of the farmers changed the technology; although the principle of diffused-light storage caught on, it was modified on each farm according to the local conditions, the household architecture and the budgets of the farmers.

An advantage of such experience is that it can help scientists and extension workers to better understand local conditions. There has long been a tradition, aptly named by Robert Chalmers, of IDS, as 'rural development tourism', in which periodic visits to the field are seen as providing sufficient 'feel' for conditions [18]. But typically such visits are confined to the accessible roadside areas and to meetings with headmen and local experts and are usually conducted at those times of the year when travel is easiest. They inevitably produce biased impressions which can be seriously misleading. Much rural poverty remains unperceived. Such visits are no substitute for intimate and systematic analyses of the circumstances and livelihoods of rural households.

In 1978 Ian Craig and I and a team of natural and social scientists at the University of Chiang Mai in northern Thailand, led by Benjawan and Kanok Rerkasem, Manu Seetisarn and Phrek Gypmantasiri, developed a technique to address this need which goes under the name of Agroecosystem Analysis (AEA) [19]. A Ford Foundation grant had been given to the university in 1968 to create a Multiple Cropping Project (MCP) aimed at designing advanced, triple-crop rotational systems with which farmers could capitalize on the government irrigation schemes recently installed in the Chiang Mai valley. The foundation also gave many of the young staff scholarships to go abroad for further graduate training. They returned in the late 1970s, eager to put their new skills and experience to the task of helping the farmers of the valley. But, as they soon became aware, much of the work of the MCP in the intervening years had not proved particularly relevant. Although the MCP had developed some half dozen, apparently superior and productive, crop systems there were very few cases of adoption by the farmers; on the other hand, the farmers themselves had developed a large number of triple-crop systems in response to the new opportunities provided by the irrigation systems.

This raised questions in their minds as to the role they, as university researchers, could most effectively play. In terms of helping the farmers of the valley, where did their comparative advantage lie? Should they continue to design new systems; if not, what kind of research should they undertake? These questions, they realized, could not be answered until they had a better idea of the existing farming systems in the valley and the particular problems the farmers were facing.

AEA begins with a series of extended field trips, employing direct observation and interviews with farmers to produce a set of maps and other simple diagrams which summarize the patterns of events and activities in the local agroecosystems. The diagrams are then used, in intensive, multidisciplinary workshops, to identify and discuss the key agroecosystem issues which I described in the previous chapter: productivity, stability, sustainability and equitability. Out of the workshops come a set of key questions and hypotheses requiring further investigation and research. For example, the maps in Figure 10.1, produced as part of the AEA of the Chiang Mai valley, revealed that the intensive triple-cropping developed by the farmers was largely confined to those areas under traditional or mixed (traditional plus government) irrigation systems, rather than the areas supplied by the government (RID) systems alone. Farmers found they could not rely on the government irrigation supplies. A key research question for the group was how to improve the control and reliability of irrigation in the valley. In the fifteen years since AEA was developed by the Chiang Mai group, key questions such as this have provided a basis for a very successful programme of research targeted at the needs of the farmers in the valley.

Following its success at Chiang Mai, AEA was taken to Khon Kaen University in the north-east of Thailand, where it was used to analyse the problems of the semi-arid agroecosystems of the region [20], and thence to Indonesia, where it was applied, respectively, to the uplands of East Java, the tidal swamplands of Kalimantan and the semi-arid drylands of Timor [21]. Although originally a technique designed for university and research-station workers, subsequent applications in both developing and developed countries have been designed to meet the needs of government agencies, extension workers and non-governmental organizations [22]. As the technique has spread it has evolved; the original repertoire of diagrams has been added to, as individuals have developed new ways of representing their observations and findings. Transects (Figure 10.2), map overlays, seasonal calendars, impact flow diagrams, Venn diagrams, preference rankings, decision trees – to name a few – now constitute a rich array of tools for analysis. Experience has increasingly shown the power of simple diagrams to generate productive discussion among researchers from different disciplinary backgrounds and, most significantly, to stimulate genuine interchange between researchers and farmers. For example, the tree preference ranking in Box 10.2 produced a valuable dialogue between the foresters and the farmers and helped identify what other species would be useful introductions, one being a tree with better furniture-building properties.

An illustration of the power of the diagrammatic techniques of AEA was their use in resolving a conflict over the construction of a dam on Lake Buhi in the Philippine province of Luzon [23]. The dam, which was

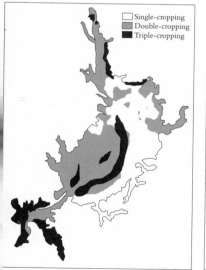
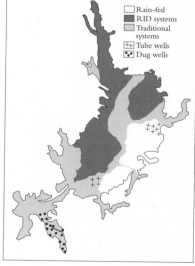

Figure 10.1 Sketch maps showing patterns of cropping intensity (a) and irrigation systems (b) in the Chiang Mai valley, northern Thailand [24]

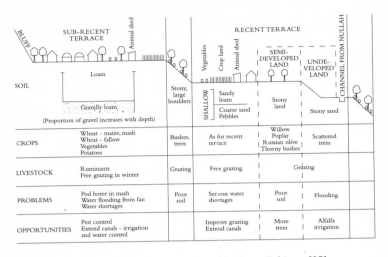

Figure 10.2 A transect of a village in Hunza, Northern Pakistan [25]

intended to provide water for a downstream irrigation system of some 10,000 hectares, was not in operation when I visited it in 1985. The fishermen above the dam were complaining of their fish cages drying out and lakeside farmers of their ricefields suffering drought. The situation had become tense; with the villagers gaining support from the New People's Army and armed village guards on the dam a stalemate had been created. With a team from the University of the Philippines in Los Banos, led by Percy Sajise, we spent two weeks conducting an AEA which culminated in a workshop attended by some seventy representatives of the aid agencies, local politicians and leaders of the farmers and fishermen. In resolving the conflict one of the AEA diagrams, a seasonal calendar (Figure 10.3), proved crucial. It pinpointed the critical constraints on the timing of different operations: the pattern of rainfall affected the possible lake levels which, in turn, determined the optimal times for both the irrigated rice and the rice planted on the drawdown around the lake shore. Typhoons and sulphur upwellings (*kanuba*) also affected rice production but, more important, limited both capture fishery and the operation of the fish cages. The outcome of the workshop was a decision to retain the water in the lake above a critical level to the end of May. This was accepted as satisfactory to upstream farmers and fisherman, as well as to the farmers downstream of the dam.

This and similar exercises were primarily designed for experts: researchers, extension workers and aid officials. The analyses, although deliberately aimed at understanding the ecological and socio-economic nature of local agroecosystems, were still designed and driven in a top-down fashion. However, the increasing involvement of farmers, as in the Buhi

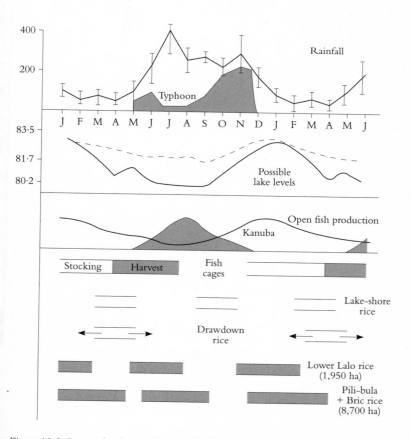

Figure 10.3 Seasonal calendar for Lake Buhi, Philippines [26]

project, not just as sources of information but as participants in analysis, began to suggest to a number of us that a more revolutionary approach was possible. In 1987 I took part with Robert Chambers, Jenny McCracken of IIED, and Constantine Berhe of the Ethiopian Red Cross in an analysis of two villages in the Wollo province of Ethiopia, site of severe drought and related mortality only four years previously. The teams of analysts included government officials, agricultural and forest experts, Red Cross activists and local leaders. Although it began as a standard AEA exercise Robert Chambers brought to the process experience from Rapid Rural Appraisal (RRA), an approach developed in the 1970s of which he was one of the pioneers.

Although RRA, like AEA, is a method of extracting information from rural people it is more informal in style, relying on semi-structured interviewing, participant observation, gaming and extended discussions [27]. It was developed, in part, as a reaction to the standard questionnaire approach of much of rural development analysis, in which fixed sets of questions are asked of rural people, the interviewers often being hired for the job and confining themselves to obtaining answers, whether accurate or not. By contrast, semi-structured interviewing takes place in the normal surroundings of the interview, does not involve a written questionnaire and, while some of the questions are predetermined, new questions or new lines of questioning arise as the interview proceeds. The intention is to create genuine dialogue in which experience and knowledge are compared and exchanged. Sometimes individuals are interviewed, on other occasions groups of people are brought together: village leaders, key informants, rich farmers, poor farmers, women, the elderly.

In Wollo we began to combine semi-structured interviewing and diagram-making, and in the process we became even more aware of the richness of analytical understanding that is common, if not normal, in rural people in the developing countries. The forest-tree rankings, described earlier, revealed not only knowledge but a capacity to make choices which balanced a wide range of considerations. At the end of the exercise we undertook a group assessment of the various options open to the inhabitants of the two villages and produced rankings that significantly shifted the priorities of the government officials present, which had previously focused on irrigation and reforestation (Table 10.1).

Robert Chambers came away from the Wollo exercise convinced it was possible to move from the extractive mode of AEA and conventional

Table 10.1 Innovation assessment in Wollo province, Ethiopia [28]

Option	Productivity	Stability	Sustainability	Equity	Cost	Time to benefit	Technical feasibility	Social feasibility
Reforestation	+	++	++	○	□	□	■	◪
Agroforestry	+	++	++	++	◪	◪	■	◪
Home gardens	++	++	++	○	■	◪	■	■
Small-scale irrigation	+	+	+	+	◪	◪	◪	■
Short-cycle varieties	++	++	++	–	◪	◪	◪	◪
Credit	○	++	++	++	◪	◪	□	■

– Negative impact
○ No impact
+ Positive impact
++ Very positive impact

	Cost	Time	Feasibility
□	High	Long	Low
◪	Medium	Medium	Medium
■	Low	Short	High

RRA to an approach in which rural people took the lead. In the ensuing months, experiments in villages in several countries demonstrated the capacity of rural people to produce their own diagrams, often in ways that showed great ingenuity and depth of knowledge. Maps, it was discovered, were readily created by providing villagers with chalk and coloured powder and no further instruction other than the request to produce a map – of the village, or the watershed or of a farm. A threshing floor or a cleared space in the village square was sufficient for villagers to produce excellent maps, often of considerable complexity. Sometimes the maps turned into models, and I remember one such in India where the farmers had collected soil from the various fields in the village and constructed a representation that faithfully laid out each field using its own soil. Maps, however, were only a beginning. It was found that seasonal calendars could be constructed by people who were illiterate and barely numerate, using pebbles or seeds. A row of twelve pebbles is laid out on the ground to indicate the months of the year and then for various items – rainfall, availability of food, labour demand, risk of illness – the relative level of activity in each month is indicated by placing a number of seeds next to each pebble.

This new approach was rapidly taken up with enthusiasm, particularly by leaders of NGOs, who were eager to find ways of creating greater levels of participation. Individuals such as Sam Joseph of Action Aid, Jimmy Marcarenhas and Aloysius Fernandez of MYRADA (Mysore Relief and Development Agency), and Parmesh Shah of the Aga Khan Rural Support Programme continued the experiments, training their own staff to facilitate the exercises. The range of diagrams quickly expanded; all those originally developed under AEA were accessible to farmers and new ones were added. Farmers, it was found, could construct pie diagrams – pieces of straw and coloured powder laid out on an earthen floor were used to indicate relative sources of income. In the process villagers have often revealed more complex and detailed knowledge than would have occurred through discussion or simple question–and–answer sessions [29]. In Robert Chambers's words, the diagrams 'have often astonished scientists, and farmers themselves, with the detail, complexity and utility of information, insight and assessment they reveal' [30].

Although this was in itself encouraging, the power of the approach was soon revealed in the use to which the diagrams could be put. Farmers were encouraged to use scoring matrices to evaluate different crop varieties in breeding programmes [31]. Maps, seasonal and pie diagrams not only revealed existing patterns, but pointed to problems and opportunities, and were seized on by rural people as a means of making their needs felt and as a basis for collective planning. The relationship between 'expert outsiders' and village people began to change. Productive dialogues replaced the traditional one-way flow of information and instruction. I remember a group of villagers in Haryana, north-western India, constructing a map of their watershed on the ground, and using four colours to indicate the degrees of degradation they observed. Present was a conservator of forests for the state, noting and comparing this classification with that of his department, which only recognized three classes. But the most enduring image was of the conservator and the villagers, down on their haunches on the ground, engaged in lively discussion about the watershed and what should be done. It was a liberating experience both for the villagers and the conservator.

The approach has now spread to most countries of the developing world, and been adopted by government agencies, by research centres and university workers as well as by NGOs. As a deliberate policy, no central manual has been produced although much has been written and there is an extensive network of practitioners. The methodologies, which are described by a bewildering variety of names – Participatory Rural Appraisal (PRA), Participatory Analysis and Learning Methods (PALM), Méthode Accélérée de Recherche Participative (MARP), to list only a few – have evolved according to local needs and customs, and reflecting local ingenuity [32]. PRA notes, produced by the IIED in London and distributed to several thousand individuals, disseminate good practice and new ideas, so that innovations in the approach reported from an African village are being tried out in an Asian village only a few weeks later.

In some ways it has been a revolution: a set of methodologies, an attitude and a way of working which has finally challenged the traditional top-down process that has characterized so much development work (Box 10.4). Participants from NGOs, government agencies and the research centres rapidly find themselves, usually unexpectedly, listening as much as talking, experiencing close to first hand the conditions of life in poor households and changing their perceptions about the kinds of interventions and research that are required [33].

Participatory approaches to development are not a recent phenomenon, but they have become increasingly common. Well known is the experience of World Neighbours in Guinope in Honduras, which has worked in partnership with the Ministry of Natural Resources and a Honduran NGO, Accorde [34]. Initially maize yields in the project area were very low (400 kg/ha), poverty and malnutrition were widespread and out-migration common. The programme started slowly and on a small scale, involving the local people in experiments with chicken manure and green manures, contour grass barriers, rock walls and drainage ditches. Extensionists were selected from the most adept farmers, and they progressively involved others so that eventually several thousand farmers participated. Maize yields tripled, and the farmers began to diversify into coffee, oranges and vegetables. Labour wages have risen from $2 to $3 a day and out-migration has been replaced by in-migration, people moving back from the slums to the homes and land they had abandoned.

Other more recent programmes have begun to make explicit use of the new analytical tools. An example is the Aga Khan Rural Support Programme (AKRSP), with whom I worked in the 1980s, in the Hunza valley and neighbouring valleys of northern Pakistan [36]. An arid mountain region, it is not naturally well endowed but the inhabitants are highly skilled in the use of the local natural resources. The development

---

Box 10.4 Comparison of top-down 'transfer–of–technology' and bottom–up 'farmer–first' approaches to development [35]

	Transfer-of-technology	*Farmer-first*
Farming conditions to which applied or more applicable	Simple, uniform, controlled	Complex, diverse, risk-prone
Main objective	Transfer technology	Empower farmers
Analysis of needs and priorities	Outsiders	Farmers facilitated by outsiders or other farmers
Transferred by outsiders to farmers	Precepts, messages, package of practices	Principles, methods, basket of choices
The 'menu'	Fixed	A la carte
Farmers' behaviour	Hear messages, act on precepts, adopt, adapt or reject package	Apply principles, use methods, choose from basket and experiment
Outsiders' desired outcomes	Widespread adoption of package	Wider farmers' choice and enhanced benefits and adaptability
Roles of outsider	Teacher, trainer, supervisor, service provider	Convenor, facilitator, consultant, searcher for and provider of choice

---

programme consists of a series of interactive dialogues through which the villagers, acting as a community, identify, plan and implement a key infrastructure project in each village. In many cases the projects are impressive feats of engineering that bring irrigation water from the glaciers to open up new land for agriculture. Several hundred such projects have been completed and the programme is now engaged in realizing the potential of the new infrastructure through a variety of initiatives funded by the villagers' savings and using diagramming techniques to determine options for innovation.

Another programme, facilitated by AKRSP (India), has made explicit use of PRA techniques in developing soil and water conservation in Gujarat in western India [37]. In the first stage the villagers produce maps of their watersheds, detailing the problem areas, planning appropriate soil and conservation works and choosing trees for planting using a technique of group ranking. This process takes one to six months. Next, village institutions are formed. They nominate extension volunteers, paid by the villagers, who are given training in PRA methods and in the necessary technical skills and project preparation and accounting procedures. They are responsible for managing teams of individuals who then implement the plans. Yields have grown by 20–50 per cent, yet the costs of the watershed treatment are 1,340 rupees/ha compared with 3,000–7,000 rupees on nearby government programmes.

In subsequent chapters I illustrate how similar participatory approaches to research and development are being applied to such diverse problems as integrated pest management systems, nutrient enhancement, the construction and management of small-scale irrigation, and reforestation and the conservation of watersheds.

Notes

[1] R. Chambers, N. C. Saxena and T. Shah, 1989, *To the Hands of the Poor: Water and Trees*, New Delhi (India), Oxford and IBH Publishing Co.

[2] G. R. Conway, 1997, 'Practical innovation: partnerships between scientists and farmers', in J. C. Waterlow, D. G. Armstrong, L. Fowden and R. Riley (eds.), *Feeding a World Population of More Than Eight Million People: A Challenge to Science*, Oxford, Oxford University Press

[3] W. D. Hooper and H. B. Ash, 1935, *Marcus Porcius Cato on Agriculture. Marcus Terentius Varro on Agriculture*, Cambridge, Mass., Harvard University Press/London, Heinemann (Loeb Classical Library)

[4] J. N. Pretty, 1995, *Regenerating Agriculture: Policies and Practice for Sustainability and Self-reliance*, London, Earthscan

[5] C. Reij, I. Scoones and C. Toulmin (eds.), 1996, *Sustaining the Soil: Indigenous Soil and Water Conservation in Africa*, London, Earthscan; P. Richards, 1985, *Indigenous Agricultural Revolution*, London, Hutchinson; I. Scoones and J. Thompson (eds.), 1994, *Beyond Farmer First: Rural People's Knowledge, Agricultural Research and Extension Practice*, London, Intermediate Technology

[6] S. M. Conway and G. R. Conway (in press), *Lanna: A Million Ricefields*, Chiang Mai (Thailand), Silkworm Books

[7] J. A. Ashby, C. A. Quiros and Y. M. Rivera, 1987, 'Farmer participation in on-farm trials', *Agricultural Administration (Research and Extension) Network*, London, Overseas Development Institute (Discussion Paper 22)

[8] Ethiopian Red Cross, 1988, *Rapid Rural Appraisal: A Closer Look at Rural Life in Wollo*, Addis Ababa, Ethiopian Red Cross Society/London, International Institute for Environment and Development

[9] Ethiopian Red Cross, op. cit.

[10] L. Sperling and U. Scheidegger, 1995, *Participatory Selection of Beans in Rwanda: Results, Methods and Institutional Issues*, London, International Institute for Environment and Development (Gatekeeper Series No. 51)

[11] Ashby *et al.*, op. cit.

[12] Sperling and Scheidegger, op. cit.

[13] J. R. Witcombe, A. Joshi, K. D. Joshi and B. R. Sthapit, 1996, 'Farmer participatory crop improvement, 1. Varietal selection and breeding methods and their impact on biodiversity', *Experimental Agriculture*, 32, 445–60

[14] Sperling and Scheidegger, op. cit.

[15] Ashby *et al.*, op. cit.

[16] D. Millar, 1994, 'Experimenting farmers in northern Ghana', in Scoones and Thompson, op. cit., pp. 160–65

[17] R. Rhoades, 1987, 'Farmers and experimentation', *Agricultural Administration (R and E)*, London, Overseas Development Institute (Network Paper 21); R. Rhoades and R. Booth, 1982, 'Farmer-back-to-farmer: a model for generating acceptable agricultural technology', *Agricultural Administration*, 11, 127–37

[18] R. Chambers, 1983, *Rural Development; Putting the Last First*, Harlow (UK), Longman

[19] G. R. Conway, 1985, 'Agroecosystem analysis', *Agricultural Administration*, 20, 31–55; 1990, 'Participatory analysis for sustainable agricultural development', *Special Lectures on the Occasion of the 25th Anniversary of the Faculty of Agriculture*, Chiang Mai (Thailand), Faculty of Agriculture, University of Chiang Mai; P. Gypmantasiri, A. Wiboonpongse, B. Rerkasem, I. Craig, K. Rerkasem, L. Ganjanapan, M. Titayawan, M. Seetisarn, P. Thani, R. Jaisaard, S. Ongprasert, T. Radnachaless and G. R. Conway, 1980, *An Interdisciplinary Perspective of Cropping Systems in the Chiang Mai Valley: Key Questions for Research*, Chiang Mai (Thailand), Faculty of Agriculture, University of Chiang Mai

[20] KKU–Ford Cropping Systems Project, 1982, *An Agroecosystem Analysis of Northeast Thailand*, and 1982, *Tambon and Village Agricultural Systems in Northeast Thailand*, Khon Kaen (Thailand), Faculty of Agriculture, Khon Kaen University

[21] KEPAS, 1985, *The Critical Uplands of Eastern Java: An Agroecosystem Analysis*; 1985, *Swampland Agroecosystems of Southern Kalimantan*; 1986, *Agro-ekosistem Daerah Kering di Nusa Tenggara Timur*, Jakarta (Indonesia), Agency for Agricultural Research and Development (KEPAS)

[22] G. R. Conway, 1986, *Agroecosystem Analysis for Research and Development*, Bangkok (Thailand), Winrock International

[23] G. R. Conway and P. E. Sajise, 1986, *The Agroecosystems of Buhi: Problems and Opportunities*, Los Banos (Philippines), Program on Environmental Science and Management, University of the Philippines; G. R. Conway, P. E. Sajise and W. Knowland, 1989, 'Lake Buhi: resolving conflicts in a Philippine development project', *Ambio*, 18, 128−35

[24] Gypmantasiri *et al.*, op. cit.

[25] G. R. Conway, Z. Alam, T. Husain and M. A. Mian, 1985, *An Agroecosystem Analysis for the Northern Areas of Pakistan*, Gilgit (Pakistan), Aga Khan Rural Support Programme

[26] Conway *et al.*, 1989, op. cit.

[27] I. Carruthers and R. Chambers, 1981, *Rapid Rural Appraisal: Rationale and Repertoire*, Brighton (UK), Institute for Development Studies, University of Sussex (IDS Discussion Paper, 155); G. R. Conway and J. A. McCracken, 1990, 'Rapid Rural Appraisal and Agroecosystem Analysis', in M. A. Altieri and S. B. Hecht (eds.), *Agroecology and Small Farm Development*, Florida, CRC Press; Khon Kaen University, 1987, *Proceedings of the International Conference on Rapid Rural Appraisal, 2−5 September*, Khon Kaen (Thailand), Khon Kaen University

[28] Ethiopian Red Cross, op. cit.

[29] R. Chambers, 1997, *Whose Reality Counts? Putting the Last First*, London, Intermediate Technology

[30] R. Chambers, 1996, *Behaviour and Attitudes: a Missing Link in Agricultural Science?*, paper presented at the second International Crop Science Congress, New Delhi, 17−24 November 1996

[31] M. Drinkwater, 1993, 'Sorting fact from opinion: the use of direct matrix to evaluate finger millet varieties', *RRA Notes*, 17, 24−8; M. Manoharan, K. Velayudham and N. Shunmugavalli, 1993, 'PRA: an approach to felt needs of crop varieties', *RRA Notes*, 18, 66−8; The Women of Sangams Pastapur, Medak, Andhra Pradesh, and M. Pimbert, 1991, 'Farmer participation in on-farm varietal trials: multilocational testing under resource-poor conditions', *RRA Notes* (London, International Institute for Environment and Development), 10, 3−8

[32] Chambers, 1997, op. cit.; A. Cornwall, I. Gujit and A. Welbourn, 1994, 'Acknowledging process: methodological challenges for agricultural research and extension', in Scoones and Thompson, op. cit., pp. 98−117

[33] IDS, 1996, *The Power of Participation: PRA and Policy*, Brighton (UK), Institute of Development Studies, University of Sussex (IDS Policy Briefing, Issue 7)

[34] R. Bunch, 1983, *Two Ears of Corn: A Guide to People-centred Agricultural Improvement*, Oklahoma City, Okla., World Neighbours; 1989, 'Encouraging farmer's experiments', in R. Chambers, A. Pacey and L.-A. Thrupp (eds.), *Farmer*

203

First: Farmer Innovation and Agricultural Research, London, Intermediate Technology; Pretty, op. cit.

[35] Chambers, 1997, op. cit.

[36] G. R. Conway, 1988, 'Rapid Rural Appraisal for sustainable development: experiences from the northern areas of Pakistan', in C. Conroy and M. Litvinoff (eds.), *The Greening of Aid*, London, Earthscan; Conway *et al.*, 1985, op. cit.

[37] Pretty, op. cit.; P. Shah, 1994, 'Participatory Watershed Management in India: the experience of the Aga Khan Rural Support Programme', in Scoones and Thompson, op. cit., pp. 117–23

11   Controlling Pests

> Creative and ambitious measures must be taken to shatter the deeply ingrained uncritical and dependent attitude towards pesticides which prevails at all levels in developing countries, from ministries to the smallest farms.
>
> — Patricia Matteson, Kevin Gallagher and Peter Kenmore,
> 'Extension of integrated pest management for planthoppers in Asian irrigated rice: empowering the user' [1]

Pests, pathogens and weeds are the most visible of threats to sustainable food production [2]. Just how much crop and livestock loss they cause is largely guesswork; estimates range between 10 and 40 per cent. But in some situations the potential losses can be considerably higher. Much depends on the nature of the crop: where a premium is placed on the quality of the harvested product – for example, cotton, or fruits or vegetables – even a small pest or pathogen population can cause the farmer serious financial loss. Grain crops are not in this category, but the intensity of cultivation of the new varieties encourages severe pest and pathogen attack, on occasion resulting in total destruction of the crop.

Since the Second World War, the common approach to pest, pathogen and weed problems has been to spray crops with pesticides (insecticides, nematocides, fungicides, bactericides and herbicides). Quite apart from the hazards they pose to human health and wildlife (Chapter 6), they are frequently costly and inefficient. This has been especially true of the modern insecticides: they have to be repeatedly sprayed if control is to be maintained, insect pests commonly become resistant to them and, as ecological research has shown, they can make the problem worse by killing off the natural enemies – the parasites and predators – which normally control pests [3].

I first encountered the problems pesticides can cause when I was working as an ecologist in North Borneo (later the State of Sabah, Malaysia) in

1961. Cocoa was a recently introduced crop being grown in large, partial clearings in the primary forest. When I arrived the crop was being devastated by pests: cocoa loopers and bagworms were stripping off all the leaves, cossid borers were destroying the branches, ring-bark borers were killing whole trees, and a pest new to science, the bee bug, was damaging the cocoa pods [4]. At the time the cocoa fields were being heavily and repeatedly sprayed with insecticides, sometimes consisting of cocktails of organochlorines, such as DDT and dieldrin. They were having little effect. On the contrary, I believed they were making the situation worse. In their natural forest home the pest species were probably being controlled by a variety of natural enemies and, it seemed to me, the problem was being caused by the pesticides, which, being unselective in their action, were killing off the natural enemies.

At my recommendation all spraying was stopped. Two of the pests, the branch borer and the cocoa looper, soon came under control by parasitic wasps. The bagworms continued to cause damage and they were controlled by use of a highly selective pesticide before eventually being naturally controlled by a parasitic fly. The ring-bark borer was largely eliminated by destroying a secondary forest tree that had remained in the fields and was the borer's natural host. Very selective spraying kept the bee bug in check. Within a year all the major pests were being satisfactorily controlled, and this has persisted to the present day (Figure 11.1).

Since the 1960s the broad-spectrum organochlorine insecticides have

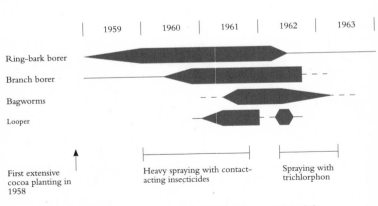

Figure 11.1 The control of cocoa pests in North Borneo, 1961 [5]

been replaced by more selective compounds, which also tend to be less damaging to wildlife and human health [6]. Increasingly stringent regulations in the developed countries have forced manufacturers to engage in exhaustive safety and environmental testing, both in the laboratory and in natural field conditions [7]. New pesticides were discovered in the past by a largely random process of screening thousands of synthetic compounds. Now, with the greater understanding conferred by modern cellular and molecular biology, chemical companies have begun to search for tailor-made pesticides. One group are compounds that mimic the effects of juvenile hormones in insects, disrupting the transition from one lifecycle stage of an insect to another, for example preventing caterpillars from becoming moths. They are valuable because they often only affect one species of insect. Another successful group of insecticides is based on the bacterium, *Bacillus thuringiensis*. When ingested by a caterpillar feeding on a sprayed leaf, a toxic protein is released which paralyses the caterpillar's gut and mouthparts, causing it to die. Since natural parasites and predators do not feed on the sprayed leaves they are unaffected. As I recounted in Chapter 8, this property of *Bacillus thuringiensis* has been exploited in genetic engineering.

There is also growing interest in natural plant compounds that have been traditionally used by farmers for pest control. They include custard apple, turmeric, croton oil tree, Simson weed, castor oil, ryania and chilli pepper [8]. The pyrethroids, based on the compound pyrethrum found in chrysanthemum plants, are effective against certain pests and are very safe. One of the best-known sources of a natural insecticide is the neem tree, which has been used against rice pests in India for centuries [9]. The bitter compound, azardirachtin, contained in the seed acts as an anti-feedant, making crops unpalatable to pests. It does not harm birds or mammals or beneficial insects such as honey bees. Unfortunately, it degrades fairly rapidly in sunlight. An effective formulation which prevents azardirachtin from degrading is on the market, but is considerably more costly than the natural product [10].

An alternative, or a complement, to using selective pesticides is to encourage the natural enemies of pest directly. Sometimes, although rarely, this can be spectacularly successful. A recent example is the biological control of the cassava mealybug in Africa [11]. The mealybug first appeared in the Congo and Zaire in 1973, but soon spread across a wide belt of central Africa, from Mozambique to Senegal, producing yield losses of up to 80 per cent. Cassava originated in South America and a search was made there for the mealybug's natural enemies. A parasitic wasp was found in Paraguay and released in Nigeria in 1981. The results were dramatic, with yield increases of up to 2·5 tons/ha; overall benefits are estimated in terms of billions of dollars. Plant pathogens can also be controlled by their 'enemies', organisms which act as antagonists [12]. In the United States, commercially available *Agrobacterium radiobacter* (K84) produces an anti-biotic which prevents the growth of the pathogen causing crown galls [13]. Biological control has also sometimes been effective against weeds, by releasing herbivorous insects – such as leaf-eating beetles. But more promising is the use of the phenomenon of 'allelopathy'. Certain plants release compounds which are harmful to weeds. One approach may be to introduce the allelopathic genes into crop plants; another may be to synthesize the toxic allelopathic compounds [14].

Often natural enemies of pests can be encouraged by creating a more diverse agroecosystem. There are very few pest problems in the Javanese home gardens described in Chapter 9. The diversity of plants in each garden encourages a diversity of insects which, in turn, supports a large population of general predators – spiders, ants, assassin bugs – that keep potential pests under control. Sometimes even growing a mixture of two crops is enough [15]. In the Philippines, intercropping of maize and peanuts helps to control the maize-stem borer. The predator is a spider which, as an adult, feeds on the stem borer caterpillars. But the young spiders feed on springtails and these they find in the leaf litter under the peanut plants. The simple intercrop is sufficient to create a complex and beneficial food web. Often the mechanism of control is subtle. Aromatic odours from the intercropping of cabbages and tomatoes repel the diamondback moth; the shading effect of mung bean or sweet potato grown with maize reduces weed growth; and the liberation and spread of pathogen inocula can be reduced by growing cowpeas with maize [16].

The move towards large areas of monoculture has been one of the reasons why pest and disease outbreaks have grown in the wake of the Green Revolution [17]. There have been other factors. Rice-stem borer and sheath blight attacks have increased as a result of higher nitrogen applications and leaf disease is more prevalent in the microclimate created by the densely leaved, short-strawed wheats and rices (although, it should be noted, fertilizers increase resistance to rice tungro virus, while irrigation reduces losses to rice blast) [18]. The narrow genetic stock of the new varieties has also been a contributory factor, as has the misuse of pesticides.

Pests and pathogens, in common with all organisms, have the ability to adapt, through natural selection, to new situations. Michael Loevinsohn, working in the Philippines, has shown the remarkable capacity of rice pests to evolve in response to variation in the timing of rice cultivation. Within a few years genetically different populations of pests have arisen, each adapted to rice-cropping patterns separated by only a few kilometres. At Mapalad, at the base of the Sierra Madre Mountains in Luzon, where a single rain-fed crop is grown, populations of the yellow stem borer had a shorter generation time than did populations at Zaragoza, 10 km away in the centre of the irrigated plain where two crops are grown. They also laid more eggs and had a lower survival rate. Planting is carried out more or less at the same time at Mapalad and the crop matures uniformly; in these conditions there is a selective advantage for pests that mature quickly and increase rapidly in numbers. By contrast, under irrigated double-cropping the planting is asynchronous and the pests are more heavily attacked by natural enemies – predators and parasites. In these circumstances it is advantageous, for the pests, to mature more slowly but have a higher survival rate.

Not surprisingly, pest and pathogen populations responded very rapidly to the continuous cropping of the new varieties of wheat and rice. The first ten years of double-cropping of rice at IRRI in the Philippines resulted in dramatic growth in pest populations (Figure 11.2). The numbers increased directly because of the introduction of a dry-season rice crop; but there were more pests and more damage on the wet season crop as well. Thirteen per cent of the wet season crop was lost under single-cropping, but this rose to 33 per cent when double-cropping was introduced [19]. Under triple-cropping the numbers and damage were even higher. Only where there is a break in cultivation, such as a fallow, or where a cereal is alternated with a dissimilar crop, are pests and diseases held in check.

Pests and pathogens are also capable of evolving rapid resistance to threats and adverse circumstances, in particular to the use of pesticides (Figure 11.3) [20]. By the mid-1980s some 450 pest species in the world were resistant to one or more insecticides and about 150 fungi and bacteria were resistant to or tolerate fungicides. Nearly 50 weed species were resistant to herbicides. Several important insect pests are resistant to all the major classes of insecticides: the diamondback moth, a pest of cabbages and other crucifers, is resistant in Malaysia not only to the older organo-

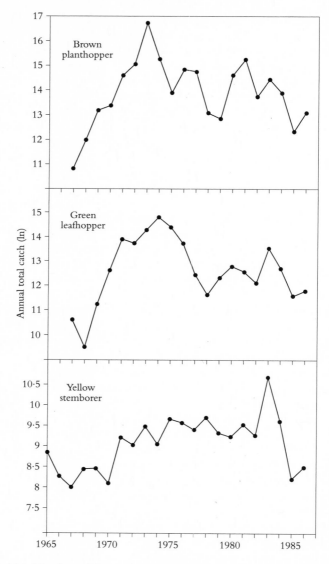

Figure 11.2 Numbers of three rice pests at IRRI, 1965–86 [21]

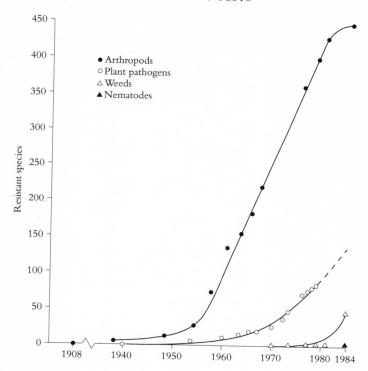

Figure 11.3 Resistance to pesticides [22]

chlorines and carbamates but also to the newer organophosphates and pyrethroids [23].

Pests and pathogens are equally adept at evolving ways of overcoming the defences which naturally occur in crop plants or are bred into them by plant breeders. In 1950 a new race of wheat-stem rust suddenly exploded in the United States and southern Canada, and was carried by high winds into Mexico [24]. This was only the first of a series of epidemics. New races continued to arrive and by 1960 a group of virulent races had almost completely replaced the existing, relatively weak forms. The wheat-breeding programme was able to keep pace with this changing pattern of disease but only by virtue of having new resistant varieties quickly available when each change in race occurred. A similar situation arose when new biotypes of the brown planthopper (BPH) suddenly appeared on rice in South-East Asia in the 1970s and 1980s (see below).

The more intense the threat to pest and disease populations, the faster they are able to evolve mechanisms to negate or avoid the threat. Because of the high value of the new varieties – the costs of the inputs and expected magnitude of the returns – attempts to protect them from pests and diseases have been very vigorous. Sometimes this has worked; on other occasions it has elicited a strong reaction and pest and disease situations have worsened.

In the early days of the Green Revolution heavy reliance was placed on the organochlorine pesticides that had come on the market in the post-war years and were known to be potent pesticides. IRRI, for example, advocated the use of BHC placed in cans in the channels between the fields. Yet the organochlorines, as my own work in North Borneo had shown, tended to kill not only pests but their natural enemies as well. And George Rothschild, in Sarawak in southern Borneo, had demonstrated that natural enemies played a key role in the control of rice pests [25]. These findings were largely ignored. Organophosphate insecticides were substituted for the organochlorines, but the problems became worse. Resistance emerged – one compound, diazinon, was so powerful in its ability to elicit resistance that it was used in pest-control experiments to create large pest populations. The most severe consequence, however, was the epidemic of the pest known as the brown planthopper (BPH), which became widespread on rice crops in South and South-East Asia causing, in Indonesia in particular, devastating losses in the 1970s and 1980s [26].

The brown planthopper is a sucking bug which, when present in large numbers, causes distinctive hopper-burn of the rice plants and loss of yields. It will also transmit viruses which attack the rice plant. BPH was virtually unknown as a pest before the introduction of the new rice varieties; at an important symposium on the major insect pests of the rice plant at the International Rice Research Institute (IRRI) in 1964 BPH was hardly mentioned. Yet it soon began to cause severe damage. By 1977 the losses in Indonesia were over a million tons of rice.

At about this time a young scientist at IRRI, Peter Kenmore, uncovered the reason why insecticides, far from controlling the pest, seemed to be associated with the outbreaks [27]. He showed that BPH is normally held in check by natural enemies in the ricefields – parasites which destroy the eggs and young nymphs and several species of wolf spider which prey on the planthopper. In North Sumatra the population density of the pests rose in direct proportion to the number of insecticide applications; farmers were treating fields six to twenty times over a four-to-eight-week period without any success [28]. Not only were the pesticides ineffective, they were being heavily subsidized by the government, at 85 per cent of their cost. Indonesia had become self-sufficient in rice production but, in the process, was accounting for 20 per cent of the world's use of pesticides on rice. In 1986, the government acted on the basis of the mounting evidence implicating pesticides in the outbreaks. A presidential decree banned 57 of the 66 pesticides used on rice and began to phase out the subsidy. Some of the savings went to fund an integrated Pest Management (IPM) programme (see below).

The principal alternative to using pesticides – the breeding of resistance to pests and diseases into plants – has had a mixed record. At IRRI there has been considerable success in progressively adding resistance to each new generation of varieties (see Table 8.1). The first variety with broad resistance was IR20, released in 1969. In some respects, it was inferior to preceding varieties; it had a rather weak stem and could not stand heavy applications of fertilizer, but its resistance and its superior grain quality made it popular in Asia for over fifteen years. Not all pests and diseases, however, are susceptible to this approach. One of the most important and persistent pests of rice is the stem borer and much effort has been devoted to producing resistant varieties, but so far with little success. The more serious defect of this strategy is the likelihood of a breakdown in resistance and the need for breeders always to keep one step ahead. IR20 eventually fell from favour because it was susceptible to the brown planthopper.

The Indonesian BPH outbreak of 1977 was initially tackled by introducing a new resistant rice variety, IR26, but within three seasons it had failed and losses in 1979 were again very severe. Next to be introduced was IR36. It was more successful, and was rapidly adopted. By 1984 Indonesia had become self-sufficient in rice production. But the resistance of IR36 was also short-lived. By 1986 the planthoppers had exploded to the levels of 1977, threatening over 50 per cent of Java's riceland. Losses in 1986/7 were estimated to be nearly US$400 million [29]. One explanation of this sequence is that BPH exists in a number of different biotypes or races:

- The original resistance in the variety IR26 had been to biotype 1;
- Then biotype 2 emerged, to which IR36 was resistant. It kept its resistance from 1977 to 1982;

– Then in 1983 a new biotype (biotype 3) invaded Sumatra and attacked IR36;
– IR56 was then introduced and was resistant to all the biotypes [30].

However, Peter Kenmore and his colleagues believe this is too simplistic an explanation [31]. In their experience, planthopper populations are extremely variable and can rapidly evolve to local circumstances (as Michael Loevinsohn has also shown for the rice-stem borer, see above). They argue that the heavy pesticide spraying accelerates the adaptation of BPH to new rice varieties. In their view plant-breeding approaches are only sustainable if they are a part of an integral strategy.

The approach plant breeders have taken in attempting to create resistant cereal varieties has been to seek out and introduce single, major resistant genes. These effectively confer immunity against one or, at most, a few strains or races of the pest or disease [32]. It has proven to be a very effective strategy. One variety, the Mexipak wheat, Sonalika, developed in Pakistan from the Mexican varieties, stood up to major rust attack in South Asia for over twenty years. But it can be risky if there is a likelihood of a new virulent pest or disease arising. Sonalika's resistance to rust is now breaking down and new races of rice blast are producing repeated breakdown of resistance in the new rice varieties [33]. Resistance to blast usually breaks down in two to three years. This is not a problem if new sources of resistance can be found quickly. When IR20, originally resistant to tungro virus, was wiped out by a new strain in the Philippines in 1972, a new resistant variety, IR26, was produced by IRRI within a year.

The alternative strategy is to build up a combination of genes each of which contributes only a partial degree of resistance. It is slower to achieve; accumulating the necessary genes from different parents can take ten to twelve years [34]. However, because the pest or pathogen is being resisted in a number of ways and is rarely completely controlled, there is less likelihood of a new, more virulent strain emerging. As experience over the past thirty years has shown, the greater the apparent success in achieving pest or disease control in the short term, the greater the likelihood of a serious breakdown. In the longer term it is better to live with low levels of pest attack, utilizing a diversity of approaches to keep on top of the problem.

Pest and pathogen control has been, and to some extent still is, a hit-and-miss affair [35]. Usually the first response is to try a pesticide; if this does not work or causes further problems, a different pesticide or an alternative method is tried. And so the process continues. Often what works for one pest on a crop will not work for another, or may actually make the other pest problem worse. Professionals in crop protection have long recognized the problem and since the 1950s have been developing a systematic approach to pest control that goes under the name of Integrated Pest Management (IPM) [36]. It looks at each crop and pest situation *as a whole* and then devises a programme that integrates the various control methods in the light of all the factors present. As practised today it combines modern technology, the application of synthetic, yet selective, pesticides and the engineering of pest resistance, with natural methods of control, including agronomic practices and the use of natural predators and parasites. As I demonstrated in one of the first applications of IPM in the developing countries, the control of cocoa pests in Sabah, the outcome is sustainable, efficient pest control that is often cheaper than the conventional use of pesticides.

A recent, highly successful example is IPM developed for the brown planthopper and other rice pests in Indonesia. Under the programme, farmers are trained to recognize and regularly monitor the pests and their natural enemies. They then use simple, yet effective, rules to determine the minimum necessary use of pesticides. The outcome is a reduction in the average number of sprayings from over four to under one per season, while yields have grown from 6 to nearly 7·5 tons per hectare. Since 1986 rice production has increased 15 per cent while pesticide use has declined 60 per cent, saving $120 million a year in subsidies. The total economic benefit to 1990 was estimated to be over $1 billion [37]. The farmers' health has improved, and a not insignificant benefit has been the return of fish to the ricefields.

In many ways, pest control is like a multidimensional game of chess. We pit ourselves against a variety of pests, drawing on a range of methods of control; the pests respond by evolving new defences. As we have learnt, it is not an unequal contest – there is rarely a final check-mate. Sustainable pest control depends on developing new strategies and tactics, in a continuing game. A useful way to envisage the contest is to characterize pests in terms of different evolutionary strategies. I have suggested there are three such strategies (recognizing, of course, that these are not conscious strategies – they have evolved through the process of natural selection) [38]:

1. *Opportunist pests*: these are the invaders, moving from place to place, multiplying rapidly, attacking many kinds of crop and causing enormous damage because of their numbers. They include locusts and army worms, and diseases such as rust.

2. *Specialist pests*: present most of the time, with low rates of increase, causing losses because they attack a very valuable part of the plant or transmit a disease – they include the rhinoceros beetle that eats the heart of the coconut tree and the green leafhopper that carries tungro disease of rice.

3. *Intermediate pests*: these lie between the two other types, but are distinguished by being controlled by natural enemies. They include the brown planthopper.

On the cocoa in North Borneo in the early 1960s, the bee bug and the ring-bark borer were specialists, the branch borer and cocoa looper intermediates and the bagworms were opportunists.

On our side we have, essentially, four methods of control at our disposal:

1. *Pesticide control*, the application of chemical compounds to directly kill or deter pests.

2. *Biological control*, the utilization of natural enemies, either by augmenting those already present or by introducing them from other regions or countries.

3. *Cultural control*, the use of agricultural or other practices to change adversely the habitat of the pest.

4. *Plant and animal resistance*, the breeding of animals and crop plants for resistance to pests.

Which control we use depends on the pests present and their strategies. The best approach is to identify one or more key pests that need tackling first. They are usually intermediate pests and need to be specifically targeted to ensure that their natural enemies are able to work effectively. For them, biological control is the appropriate strategy. Pesticides may have to be used against opportunists (e.g. the cocoa bagworms), but they should be selective, particularly if intermediate pests are present. Specialists such as the cocoa ring-bark borer can be controlled by manipulation of the crop environment. Box 11.1 provides a suggested matching between control and pest strategies.

Box 11.1 Control and pest strategies [39]

Pest/control strategy	Opportunists	Intermediates	Specialists
Pesticides	Based on forecasting	Selective	Targeted, based on monitoring
Biological control		Introduction or enhancement of natural enemies	
Cultural control	Cultivation, rotations, timing of planting		Destruction of alternative hosts
Resistance	Polygenic		Monogenic

Over the past forty years, IPM has grown into a sophisticated approach to pest control and has had a number of notable successes [40]. Savings have often been considerable. In Madagascar, a programme based on cultural control, plant resistance and moderate herbicide use has dispensed with a very costly aerial spraying programme covering 60,000 hectares of riceland [41]. Jules Pretty's review of IPM in the developing countries identified several programmes where annual savings are in the range of $1–$10 million [42]. But IPM has not been as widely adopted as might be expected. Part of the reason is that, despite its grounding in ecological principles, it has remained until recently a traditional top-down approach in its implementation. IPM programmes have been worked out by specialists and then instructions passed on to farmers.

IPM is a more complex process than one relying on a regular calendar of spraying. Farmers, it is often believed, cannot understand some of the technicalities involved. However, in recent years, this view has been effectively challenged. In Zamorano in Honduras, training programmes at the Escuela Agrícola Parameticana have been discovering what farmers do and do not know about pest control [43]. They know a great deal about bees, but are unaware of the existence of solitary wasps that prey on insects, or of parasitic wasps that, as larvae, live inside other insects. They are very knowledgeable about many aspects of the ear rot disease of maize, but not how it reproduces. They are aware that pesticides are toxic, but equate this with the smell of the pesticide, and take few precautions when they spray. In the training course they look at fungi under the microscope, they watch parasitoids emerge from pests and, in the field, observe wasps and ants preying on pests. A most rewarding result has been the farmers' readiness to experiment with their new-found understanding, integrating it with their traditional knowledge. One farmer intercropped amaranth among his vegetables to encourage predators; another placed his box of stored potatoes on an ants' nest; a third took parasite cocoons from his farm to a neighbour's farm.

The most extensive involvement of farmers in IPM has been the Indonesian rice programme which I outlined earlier [44]. By 1993 over 100,000 farmers had attended farmer field schools where they used simple Agroecosystem Analysis diagrams to understand and discuss the relationships between various pests and the rice crop. The life histories of pests and their predators and parasites are explained using an 'insect zoo' and dyes are placed in knapsack sprayers to demonstrate where the insecticide sprays end up. The schools themselves have become the basis of farmer IPM groups where farmers continue to meet to discuss their problems and to organize village-wide monitoring of pests and predator populations. In 1990 an outbreak of white-stem borer threatened to undermine the success of the programme, but the calls to revert to spraying were successfully resisted. Through the schools, farmers were taught to recognize the egg masses of the stem borers and in a massive campaign searched for and destroyed them. Only a handful of ricefields were infested a year later.

IPM in Indonesia has thus become institutionalized and hence sustainable. Since 1990 some 20 per cent of the farmer training has been paid for by the farmers themselves. Observers are convinced this accounts for the very considerable savings on pesticide applications and the attainment of higher yields. As one graduate put it, 'After following the field school I have peace of mind. Because I now know how to investigate, I am not panicked any more into using pesticides as soon as I discover some pest damage symptoms' [45]. This approach is now being extended to farmers in eight other countries of Asia.

Notes

[1] P. C. Matteson, K. D. Gallagher and P. E. Kenmore, 1992, 'Extension of integrated pest management for planthoppers in Asian irrigated rice: empowering the user', in R. F. Denno and T. J. Perfect (eds.), *Ecology and Management of Planthoppers*, London, Chapman & Hall, pp. 656–85

[2] 'Pests' include insects, mites, nematodes and vertebrate pests such as rats and Quelea birds. 'Pathogens' cause diseases and include fungi, bacteria, viruses and, in the case of livestock, various protozoa and worms. 'Weeds' are any plants that adversely compete with crop plants.

[3] G. R. Conway, 1971, 'Better methods of pest control', in W. W. Murdoch (ed.), *Environment: Resources, Pollution and Society*, Stanford, Calif., Sinauer Assoc. Inc.; D. Dent, 1991, *Insect Pest Management*, Wallingford (UK), CAB International

[4] G. R. Conway, 1972, 'Ecological aspects of pest control in Malaysia', in J. Farvar and J. Milton (eds.), *The Careless Technology: Ecological Aspects of International Development*, Garden City, NY, Natural History Press, Doubleday & Co., pp. 467–88; 1987, 'Man versus pests', in R. M. May (ed.), *Theoretical Ecology: Principles and Applications* (2nd edn), Oxford, Blackwell Scientific, pp. 356–86

[5] Conway, 1972, op. cit.

[6] G. R. Conway and J. N. Pretty, 1991, *Unwelcome Harvest: Agriculture and Pollution*, London, Earthscan; J. N. Pretty, 1995, *Regenerating Agriculture: Policies and Practice for Sustainability and Self-reliance*, London, Earthscan

[7] N. O. Crosland, 1989, 'Laboratory to experiment', *Proceedings of the Vth International Congress of Toxicology, July 1989*, Brighton (UK), pp. 184–92

[8] Pretty, op. cit.

[9] R. C. Saxena, 1987, 'Antifeedants in tropical pest management', *Insect Science and its Applications*, 8, 731–6

[10] FAO, 1993, *Harvesting Nature's Diversity*, Rome (Italy), Food and Agriculture Organization

[11] P. Neuenschwander and H. R. Herren, 1988, 'Biological control of the cassava mealybug *Phenacoccus manihoti*, by the exotic parasitoid *Epidinocarsis lopezi* in Africa', *Philosophical Transactions of the Royal Society of London*, B, 318, 319–33; A. Kiss and F. Meerman, 1991, *Integrated Pest Management in African Agriculture*, Washington DC, World Bank (Technical Paper 142, African Technical Department Series); R. Norgaard, 1988, 'The biological control of cassava mealybug in Africa', *American Journal of Agricultural Economics*, 70, 366–71

[12] R. Campbell, 1989, *Biological Control of Microbial Plant Pathogens*, Cambridge, Cambridge University Press

[13] M. A. Altieri, 1995, *Agroecology: the Science of Sustainable Agriculture* (2nd edn), Boulder, Colo., Westview Press/London, Intermediate Technology

[14] Altieri, op. cit.

[15] Conway, 1971, op. cit.

[16] Altieri, op. cit.

[17] R. F. Smith, 1972, 'The impact of the Green Revolution on plant protection in tropical and subtropical areas', *Bulletin of the Entomological Society of America*, 18, 7–14

[18] E. E. Saari and R. Wilcoxson, 1974, 'Plant and disease situation of high-yielding dwarf wheats in Asia and Africa', *Annual Review of Phytopathology*, 12, 49–68; IRRI, 1985, *Proceedings of the Second Upland Rice Conference*, Los Banos (Philippines), International Rice Research Institute

[19] M. E. Loevinsohn, J. A. Litsinger and E. A. Heinrichs, 1988, 'Rice insect pests and agricultural change', in M. K. Harris and C. E. Rogers (eds.), *The Entomology of Indigenous and Naturalized Systems in Agriculture*, Boulder, Colo., Westview Press, pp. 161–82

[20] M. Dover and B. Croft, 1984, *Getting Tough: Public Policy and the Management of Pesticide Resistance*, Washington DC, World Resources Institute; G. P. Georghiou, 1985, 'The magnitude of the problem', in National Research Council, *Pesticide Resistance: Strategies and Tactics for Management*, Washington DC, Committee on Strategies for the Management of Pesticide Resistant Pest Populations, Board of Agriculture, National Research Council, National Academy Press

[21] Annual catches of moths at light traps. – Loevinsohn, 1994, op. cit.

[22] Georghiou, op. cit.

[23] K. I. Sudderuddin, 1979, 'Insecticide resistance in agricultural pests with special reference to Malaysia', in *Proceedings of MAPPS Seminar, 1–2 March, 1979*, Kuala Lumpur (Malaysia), pp. 138–48

[24] H. Hanson, N. E. Borlaug and R. G. Anderson, 1982, *Wheat in the Third World*, Boulder, Colo., Westview Press

[25] G. H. L. Rothschild, 1971, 'The biology and ecology of rice-stem borers in Sarawak (Malaysian Borneo)', *Journal of Applied Ecology*, 8, 287–322

[26] P. Kenmore, 1991a, 'Getting policies right, keeping policies right: Indonesia's Integrated Pest Management policy, production and environment', paper presented at the Asia Region and Private Enterprise Environment and Agriculture Officers' Conference, Sri Lanka; 1991b, *How Rice Farmers Clean Up the Environment, Conserve Biodiversity, Raise More Food, Make Higher Profits: Indonesia's IPM – a Model for Asia*, Manila, Food and Agriculture Organization; K. D. Gallagher, P. E. Kenmore and K. Sogawa, 1994, 'Judicial use of insecticides deter planthopper outbreaks and extend the life of resistant varieties in Southeast Asian rice', in Denno and Perfect, op. cit., pp. 599–614; R. Stone, 1992, 'Researchers score victory over pesticides – and pests – in Asia', *Science*, 256, 5057

[27] P. E. Kenmore, 1980, 'Ecology and outbreaks of a tropical pest of the green revolution, the brown planthopper, Nilaparvata lugens. Stahl.', PhD thesis, Berkeley, Calif., University of California

[28] P. Kenmore, 1986, *Status Report on Integrated Pest Control in Rice in Indonesia with Special Reference to Conservation of Natural Enemies and the Rice Brown Planthopper (Nilaparvata lugens)*, Jakarta (Indonesia), Food and Agriculture Organization

[29] E. B. Barbier, 1987, 'Natural resources policy and economic framework', in J. Tarrant *et al.*, *Natural Resources and Environmental Management in Indonesia*, Annex 1, Jakarta (Indonesia), United States Agency for International Development (USAID)

[30] R. W. Herdt and C. Capule, 1983, *Adoption, Spread, and Production Impact of Modern Rice Varieties in Asia*, Los Banos (Philippines), International Rice Research Institute

[31] Matteson *et al.*, op. cit.

[32] N. W. Simmonds, 1981, *Principles of Crop Improvement*, Harlow (UK), Longman; J. E. Vanderplank, 1982, *Host−Pathogen Interactions in Plant Disease*, New York, NY, Academic Press

[33] E. E. Saari, 1985, 'South and South-east Asian region', in CIMMYT, *Report on Wheat Improvement, 1983*, Mexico, DF, International Maize and Wheat Improvement Center; S. H. Ou, 1977, 'Genetic defence of rice against disease', in P. R. Day (ed.), *The Genetic Basis of Epidemics in Agriculture. Annals of the New York Academy of Science*, pp. 275−86

[34] G. S. Khush, 1992, 'Selecting rice for simply inherited resistances', in H. T. Stalker and J. P. Murphy (eds.), *Plant Breeding in the 1990s: Proceedings of the Symposium on Plant Breeding in the 1990s*, Wallingford (UK), CAB International, pp. 303−22

[35] Conway, 1971, op. cit.

[36] M. L. Flint and R. van den Bosch, 1981, *Introduction to Integrated Pest Management*. New York, NY, Plenum Press; J. R. Cate and M. K. Hinkle, 1994, *Integrated Pest Management: The Path of a Paradigm*, Washington DC, National Audubon Society

[37] Kenmore, 1991a, 1991b, op. cit.

[38] Conway, 1971, op. cit.

[39] ibid.

[40] L. A. Thrupp (ed.), 1996, *New Partnerships for Sustainable Agriculture*, Washington DC, World Resources Institute

[41] A. Von Hildebrand, 1993, 'Integrated pest management in rice: the case of the paddy fields in the region of Lake Alaotra', paper presented at the East/Central/ Southern Africa Integrated Pest Management Implementation Workshop, Harare, Zimbabwe, 19−24 April, 1993

[42] Pretty, op. cit.

[43] J. W. Bentley, G. Rodrígues and A. González, 1993, 'Science and the people: Honduran campesinos and natural pest control inventions', in D. Buckles (ed.), *Gorras y Sombreros: Caminos hacia la Colaboración entre Técnicos y Campesionosia*, El

Zamarano (Honduras), Department of Crop Protection; 1994, 'Stimulating farmer experiments in non-chemical pest control in Central America', in I. Scoones and J. Thompson (eds.), *Beyond Farmer First: Rural People's Knowledge, Agricultural Research and Extension Practice*, London, Intermediate Technology, pp. 147–50

[44] Pretty, op. cit.; Kenmore, 1991, op. cit.; Matteson *et al.*, op. cit.; Y. Winarto, 1994, 'Encouraging knowledge exchange: integrated pest management in Indonesia', in Scoones and Thompson, op. cit., pp. 150–54

[45] E. van der Fliert, 1993, *Integrated Pest Management: Farmer Field Schools Generate Sustainable Practices*, Wageningen (Netherlands), Wageningen Agricultural University (WAU Paper 93–3)

12   Replacing Nutrients

> Once the natural vegetation is cleared 'the trees, cut down by the axe, cease to nourish their mother with their foliage'. However, 'we may reap greater harvests if the earth is quickened again by frequent, timely, and moderate manuring'.
>
> – Lucius Columella, *De re rustica* [1]

Lucius Columella, writing in the next century after Marcus Varro, who I quoted at the beginning of Chapter 9, was another Roman landowner who clearly understood the basis of sustainability. In natural ecosystems the nutrients cycle. They are collected from the soil by the roots of plants, contribute to the growth of stems, leaves and fruits, and when the plants die are returned to the soil as the vegetation rots. A similar cycling supports animal populations: nutrients are ingested as the animals graze on grass and other plants, are returned partly in the excreta and urine and partly when the animals eventually die and decompose (Figure 12.1).

As all farmers recognize, when plants are treated as crops and animals as livestock, the process of harvesting removes the nutrients from the ecosystem (Figure 12.2). Some soils are naturally richer in nutrients than others and can be mined, at least for a period, but eventually for all soils the lost nutrients have to be replaced. Without nutrient replacement there is no agricultural sustainability.

Until recently the only commonly available means of replacement available to the farmer was to apply composts or manures or, as the Romans recognized, to grow nitrogen-fixing legumes. This changed with the invention, at the beginning of this century, of the Haber-Bosch process for synthesizing ammonia from the atmosphere. Today, the manufacture of synthetic, inorganic fertilizers is highly efficient, at least in the developed countries, and relatively inexpensive, relying as it does on atmospheric nitrogen, and fossil fuels, such as methane and coal, for its basic ingredients. China is the world's largest manufacturer and India's production is growing

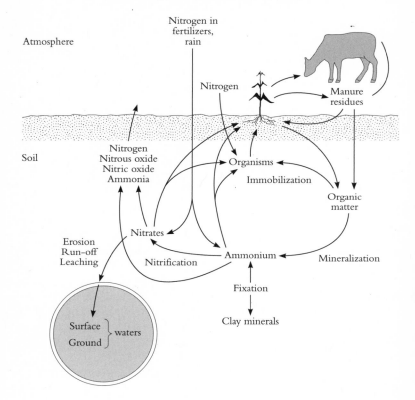

Figure 12.1 The nitrogen cycle in crops, livestock and the soil [2]

fast, although there is little capacity in Sub-Saharan Africa (Figure 12.3). The other key nutrients, phosphorus and potassium, are also fairly plentiful in mineral form. However, they are geographically narrowly confined – 50 per cent of the deposits are located in the Middle East and another 25 per cent in South Africa [3].

The development of synthetic fertilizers has opened up a potential for very high yields, far higher than were achieved by natural nutrient cycling. It was this potential that the breeding of the short-strawed rice and wheat varieties was designed to exploit. About 60 per cent of total fertilizers are today applied to cereals: over a half on rice, and a quarter on wheat [4].

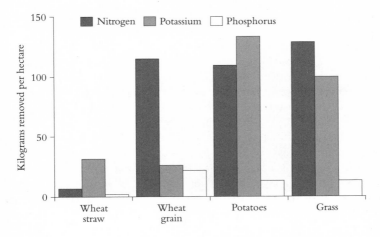

Figure 12.2 Nutrients removed by crops in the United Kingdom [5]

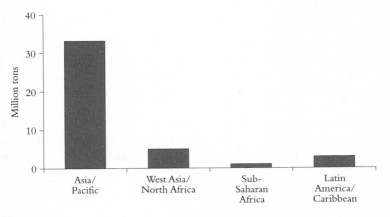

Figure 12.3 Regional production of nitrogenous fertilizers

Another 13 per cent goes on sugar cane and cotton. Very little is applied to food crops other than cereals. Recommended application rates for the new rice and wheat varieties are between 120 and 170 kg nitrogen/ha [6].

At these rates, farmers can expect returns of eight- to twentyfold in terms of kilograms of additional grain per kilogram of additional plant nutrient, and thirty- to fiftyfold for roots and tubers [7].

But, as I indicated in Chapters 6 and 7, there is a downside. The economic returns may not match the physical returns. Although world fertilizer prices have declined over the past fifteen years, in many developing countries fertilizers remain expensive. About a third of fertilizers are imported, and inefficient distribution systems put up costs and limit availability [8]. Timing of application is crucial to getting the best results and if supplies are late or erratic, farmers may incur heavy costs with little return. The total package of nutrients is also important. In addition to nitrogen (N), phosphate (P) and potassium (K), plants require a range of micronutrients. Farmers have to get the balance of nutrients right for different crops and different soils. Soil chemical degradation – excessive salinity, acidity or alkalinity – may interfere with the uptake of nutrients.

A great deal of the nitrogen fertilizer is wasted. Farmers may apply more than the optimum quantity required to maximize production or the timing may be suboptimal. This is understandable. There is massive variation in the response of crops to fertilizer applications – the soil type, the rainfall and the kind of fertilizer all affect the efficiency with which the nutrients are taken up and converted to grain or other harvested product. It is difficult, even with sophisticated analyses, to determine the appropriate level of fertilization. And, as I indicated in Chapter 6, very high levels of subsidy – 70 per cent of world prices in Indonesia in the 1980s – encourage farmers to over-fertilize.

Recovery rates – the amount of nitrogen applied that ends up in the harvested product – can be as high as 100 per cent in temperate climates. But in the tropics, even under highly controlled conditions and with the best agronomic practices, recoveries are much lower: 50–60 per cent for dryland crops but seldom more than 30–40 per cent for rice [9]. The unique anaerobic conditions of paddy fields result in heavy losses of nitrogen, particularly through volatilization as ammonia. And much of the remaining nitrogen runs off to surface waters or is leached to underground aquifers. In the seasonal tropics, leaching is encouraged by the alternation of the extremes of the wet and dry seasons. There is a slow build-up of nitrate in the topsoil during the dry season as a result of the mineralization of organic nitrogen. This is followed by a rapid and short increase at the onset of the rains and then a decline as nitrate is flushed into the surface and groundwater [10]. On large-scale irrigated ricelands, which are mostly located in the seasonal tropics and are heavily fertilized, well over half the nitrogen is lost either to the atmosphere or via the irrigation outflows [11].

Part of the answer to this waste is to improve the optimality of fertilizer application. At present, most Asian rice farmers apply fertilizer directly on to the water one to three weeks after transplanting and this results in extensive losses to the atmosphere. It would be better if the applications were split and more closely timed to the plant's need. Three applications, the first just before transplanting, the second at maximum tillering and the last just before the initiation of the flowering panicles, would be less wasteful [12]. In Europe, fields are regularly tested and farmers routinely provided with precise recommendations on application rates and timing, dependent on the soil type and the previous crop. There is an urgent need for similar advice in the developing countries.

Nitrogen utilization can also be improved by novel methods of application. Spraying the fertilizer on the leaves, for example, results in rapid absorption and translocation through the plant [13]. It cuts applications on some vegetable crops by 25 per cent. Coating urea fertilizer with sulphur produces a slower, more controlled release of nitrogen, and also reduces methane, ammonia and nitrous oxide losses [14]. But much can be done without resort to novel techniques. Losses of ammonia are greatest when fertilizers are applied to the surface of the water and can be reduced if the fertilizer is thoroughly incorporated in the soil during land preparation. Reduction of methane emissions from ricefields – as much as 88 per cent without lowering yields – can be achieved by draining the fields at specific times [15].

In addition to the 50 million tons of methane produced a year by ricefields, a further 80 million tons comes from ruminant animals, mostly cattle. In the developed countries adding antibiotics and steroids to cattle feed reduces methane emissions, as does the use of somatotrophin (BST) to increase milk production in dairy cows [16]. There are also long-term possibilities of using biotechnology to modify the process of fermentation in the cattle rumen where the methane is produced. But the most promising approach for the developing countries is to improve the diets of live-stock and the efficiency of their utilization of feed. Supplementing dairy cattle diets in India with high-quality feed reduces methane emissions per litre of milk by a factor of three [17]. An alternative is to make greater use of crop wastes and of specially grown crops such as sugar cane or the

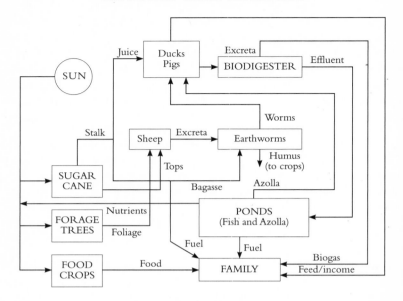

Figure 12.4 A model integrated livestock/aquaculture system for a family farm [18]

leguminous forage tree *Prosopis*. Cane juice can be fed to pigs and poultry and the forage to ruminants [19]. Various models of integrated livestock/aquaculture systems exist which have the potential to be sustainable and highly productive on a small land area and, at the same time, to reduce methane emissions from the 1 ton per ton of meat under traditional pastoral husbandry to under 100 kg (Figure 12.4).

These and many other possible approaches to reducing gaseous pollutants from agriculture have been assessed and summarized by the Intergovernmental Panel on Climate Change. Box 12.1 lists those likely to be of relevance to the developing countries. If adopted globally and in full they would achieve reductions of the related methane emissions from 130 million tons to 81 million tons and of nitrous oxide from 3·5 million tons to 2·8 million tons. But their adoption will not be straightforward. There are trade-offs. While periodic drainage of ricefields will reduce methane emissions, it may increase the emissions of nitrous oxide; and there could

---

Box 12.1 **Practices with potential for reducing gaseous emissions** [20]

METHANE REDUCTION

Ruminants
Improving diet quality and nutrient balance
Increasing food digestibility
Using antibiotics and other additives
Improved cattle breeds
Twinning to reduce number of breeding animals

Paddy fields
Drainage of fields
Sulphur-coated urea
New rice varieties

NITROUS OXIDE REDUCTION

Optimization
Soil/plant testing
Minimize fallows
Split applications
Tillage, irrigation and drainage

Tighter flows of nutrients
Integration of plants and livestock
Retention of crop residues

Novel fertilizers
Controlled release
Fertilizers below soil surface
Foliar feed application
Nitrification inhibitors
Fertilizers matched to rainfall

---

be a similar consequence from improving the feed quality of cattle. Most of the techniques in the box are likely to be costly. Market forces may stimulate appropriate research and development and eventually bring down costs, particularly where increased efficiency is involved; nevertheless this is an area where international, public-funded investment will be essential since many of the returns are likely to be societal rather than private.

Many of these techniques are more suited to the high-potential areas, but some can be adopted where the yield potential is lower. For example, incorporation of urea fertilizers, formulated as briquettes, marbles or super-granules, deep in the soil prior to planting is often more cost-effective than broadcasting the fertilizers on the surface. Use of supergranules in Taiwan increases rice yields by 20 per cent in marginal areas [21]. And trials in Indonesia with deep placement of urea have produced a 10 per cent yield increase with a 25 per cent decrease in the amount of urea applied [22]. Because deep placement requires more labour it has an advantage in the lower-potential areas where employment creation is also a priority.

In many lower-potential situations, however, synthetic fertilizers are too expensive or difficult to obtain. One answer lies in breeding crops – such as the rice variety IR42 – that will make better use of existing soil nitrogen or fix their own nitrogen. A more ambitious target is to develop an upland rice that grows in rain-fed fields, is perennial and fixes its own nitrogen (see Figure 8.1) [23]. Klaus Lampe, the former director of IRRI, referred to this as one of his 'Man on the Moon' projects. If it could be achieved it would be a truly 'miracle rice' for farmers in the lower-potential lands.

One of the biggest challenges today, particularly in Sub-Saharan Africa, is to improve the nutrient status of lands subject to bush-fallow or shifting cultivation, which I described in Chapter 9 [24]. An estimated 70 million hectares of forest land and 485 million hectares of savanna land are currently under some form of bush-fallow or shifting cultivation, supporting about 250 million people [25]. The viability of such systems depends on the relative lengths of the cropping and fallow periods. Most long-term studies of forest and savanna lands suggest sustainable, but low, yields require a cycle of more than two years cropping followed by ten years fallow. Anything less and degradation occurs. However, there is some experimental evidence that skilful management can generate sustainable, continuous high productivity on these soils. The best-known experiment was conducted by Pedro Sánchez and his colleagues at Yurimaguas in Peru on acid-infertile soils [26]. Maize has been grown in rotation with soybean for nineteen years without a diminution of yield, using appropriate levels of inorganic fertilizer and regular liming (Figure 12.5). But most long-term experiments indicate that maintaining organic matter on other than the most fertile soils is crucial [27].

Organic nitrogen in the soil can be boosted by encouraging the growth of certain micro-organisms or, more directly, by applying plant and animal manures. Several kinds of bacteria, and also certain other micro-organisms such as the blue-green algae, take up nitrogen from the atmosphere and convert it to ammonia which can be used by plants. Some are free-living in the soil, although they are often associated with the root zones of plants and their growth can be stimulated by certain crops. In the presence of the rice variety IR42 they will produce up to 40 kg of N/ha per year. However, the best practical results have come from exploiting nitrogen-fixing micro-organisms that live symbiotically in plants [28].

There is a blue-green alga, *Anabaena azollae*, living in cavities in the leaves of a small fern, azolla, that is a potentially phenomenal fixer of nitrogen – up to 400 kg N/ha per year, under experimental conditions. The fern will grow naturally in the water of ricefields without interfering with the growth of the rice plants. It quickly covers the surface, and after 100 days some 60 tons can be harvested per hectare, containing 120 kg of nitrogen. However, the nitrogen is not directly available to the rice crop; the ferns have to be incorporated in the soil. Rice yields can then be increased by a ton per hectare or more and the effect will carry over to a following crop, for example if wheat is grown after azolla-treated rice [29]. The best approach is to combine azolla with synthetic nitrogen fertilizers (Figure 12.6). Savings on fertilizers can be up to 50 per cent. In 1982 the Philippines created a National Azolla Action Programme to exploit this potential. The programme has a national inoculum centre, supporting a

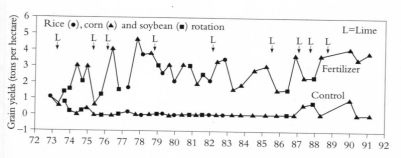

Figure 12.5 Continuous cropping on acid soils in the Peruvian Amazon [30]

231

network of regional centres, which screen and test local varieties of azolla and its blue-green alga, *Anabaena*. High-quality strains are propagated in village centres and training is provided for extension workers and farmers. The aim is to apply this approach to some 300,000 hectares of lowland rice, which would save at least $23 million, otherwise spent on imported fertilizer [31].

Best known of the symbiotic, nitrogen-fixing micro-organisms are the bacteria living in the root nodules of legumes, which can fix 100−200 kg N/ha per year. The fertilizing properties of legumes have been recognized for thousands of years. One of the earliest of the world's cropping systems − dating to soon after agriculture began in the valleys of Mexico − was the interplanting of maize and beans, often the seed of both crops being placed in the same planting hole. It is a practice which, in various forms, continues today. For example, when cowpeas are cropped together with maize, the bacteria in their root nodules can provide 30 per cent of the nitrogen taken up by the maize [32]. Cowpeas and another legume, lablab, are particularly useful in the lower-potential lands. Cowpeas are adapted to acid, infertile soils, while lablab is drought-tolerant, produces good fodder and can regrow well after clipping. Another way to capture legume nitrogen is by rotation of crops − inserting a legume such as lucerne, or a clover

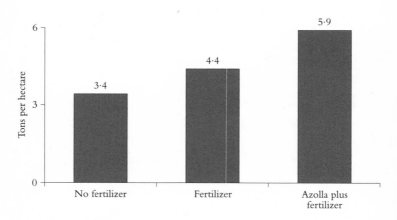

Figure 12.6 Effects of azolla and fertilizer (nitrogen, potassium and phosphorus in the ratio 30 : 30 : 30) on rice yields in the Philippines [33]

or a bean – between cereals. In the USA a variety of lucerne known as Nitro, bred for this purpose, can contribute up to 100 kg N/ha to a following maize crop [34]. There are also many bush and tree legumes that grow well in the tropics and can be interplanted with cereals and other food crops providing, under ideal conditions, 50–100 kg N/ha/year through their leaf litter or from intentional pruning (Box 12.2). By carefully timing the pruning it is often possible to ensure the nitrogen is available just when it is needed, for example to coincide with maize germination [35].

The deliberate incorporation of legume crops in the soil, known as 'green manuring', is another practice of great antiquity, yet with considerable unexploited potential today. Varro, whom I quoted at the beginning of Chapter 9, referred to some plants that are 'also to be planted not so much for the immediate return as with a view to the year later, as when cut down and left on the ground they enrich it' [36]. Most commonly used in this way was the lupin, described by Columella as first among the legumes, 'as it requires the least labour, costs least, and of all the crops that are sown is the most beneficial to the land. For it affords an excellent fertilizer for worn-out vineyards and ploughlands; it flourishes even in exhausted soil' [37]. In Bolivia today, a local lupin, *Lupinus mutabilis*, when intercropped or rotated with potatoes fixes 200 kg of Na/ha/year, minimizing the need for fertilizers and, as it happens, reducing the incidence of virus disease [38]. The best green manures in the tropics and subtropics are quick-growing legumes. In Rwanda, the shrub *Tephrosia vogelii* grows to three metres high in only ten months and produces 14 tons/ha of above-ground biomass. Maize grown after the legume has been incorporated in the soil delivers yields comparable with those achieved when the soil is heavily treated with inorganic fertilizer (Figure 12.7). Usually the best approach is to combine green manures with small amounts of inorganic fertilizer, say half the usual rate of application.

Part of the success of the World Neighbours programme in Honduras, described in Chapter 10, has been due to the successful promotion, through village extension workers, of a remarkable green manuring legume, the velvetbean. It grows rapidly, its bacteria fix large amounts (up to 150 kg) of nitrogen, and it can produce as much as 60 tons per hectare of organic matter. When grown and incorporated before maize cultivation, velvetbean can increase yields two- to threefold, to over 3 tons/ha. It can grow on most soils and, because of its active, spreading habit of growth, suppresses

---

Box 12.2 Nitrogen–fixing trees [39]

Acacia albida	High levels of soil organic matter and nitrogen beneath trees; when intercropped with unfertilized millet and groundnuts, yields are up to 100 per cent higher; Africa
Acacia tortilis	Sylvo-pastoral tree; benefits pastures and soils (like other Acacias); dense roots near surface; Africa
Calliandra calothyrus	Abundant litter with rapid decay; deep rooting; multipurpose tree; Java
Casuarina equisetifolia	Dense root mat stabilizes soil surface, especially good for sand-dune stabilization
Erythrina poeppigiana	In combination with coffee and cacoa; prunings used as mulch; Latin America
Gliricidia sepium	Potential for hedgerow intercropping
Inga jinicuil	In combination with coffee and cacao; prunings used as mulch; Latin America
Leucaena leucocephala	High biomass production; high levels of N in leaves; high root biomass
Prosopis cineraria	Benefits pastures and crops in semi-arid to dry areas; improves waterholding capacity of soil, organic matter and physical conditions
Sesbania sesban	Hedgerow intercropping; other species tolerant of waterlogging

---

weeds, so cutting labour requirements by 75 per cent [40]. Sometimes quick catch crop of a legume is sufficient. Ian Craig in north-east Thailand grew a crop of cowpeas after the first brief rains of the season and before the rice was transplanted. The cowpeas were ploughed in 45–60 days later, producing a 5–20 per cent increase in the rice yields [41].

An alternative is to grow a legume which doubles as a grain-yielding crop and a green manure. Cowpeas and lablab can be intercropped with

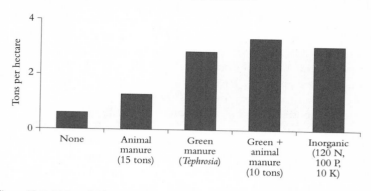

Figure 12.7 Maize yields in Rwanda under different fertilizer regimes [42]

upland rice and then, after the rice harvest, allowed to grow through the dry season. The peas are harvested and the vegetation ploughed in before the next rice crop. Yields of one ton/ha can be nearly doubled, with the added harvest of 0·5–1·0 tons of legume grain [43]. Legumes can also provide benefits in addition to increasing the availability of nitrogen and organic matter. Pigeon peas and cowpeas will access phosphate in phosphate-poor soils [44]. They are also deep-rooting, which helps water infiltration.

The other principal source of nitrogen on the farm is the livestock. If properly integrated, crops and livestock can create one of the most sustainable forms of agriculture [45]. Livestock can make efficient use, not only of purpose-grown grain and forage crops, but also of straw and the by-products of other crops. Their excreta and manure returned to the soil maintain nutrient levels and enhance soil structure. In the developed countries, there is substantial potential for utilizing manures, at least in theory: total manure production of the USA could supply some 15 per cent of the nitrogen needs of the country's farms. But the increasing geographic separation of livestock and crops creates logistic problems and raises costs. In the developing countries, there is a long, and continuing, traditional use of manures and the potential for greater utilization is more realizable. Ian Scoones and Camilla Toulmin, of IIED, describe the practice of farmers in the drylands of Africa who invite migrating pastoralists to pen their livestock overnight in the crop fields [46]. Farmers are sometimes willing to pay the herdsmen for the privilege. More recently, manures have become prized for their value in market-garden cropping. In the Chiang Mai valley of northern Thailand (see Chapter 9) 12–15 tons of manure are applied per hectare to crops such as tobacco, garlic and vegetables, grown after the traditional rice crop [47]. Farmers are often willing to pay high prices, $8–$12 per truckload in Mexico, and vegetable growers in the Guatemalan province of Quetzaltenango buy chicken wastes that are transported 100 km from Guatemala City [48].

Intensive use of livestock manures is easier if animals are penned in stalls close to the farmer's dwelling. This is becoming increasingly common in India, where the growth of dairy cooperatives is encouraging the keeping of high-yielding cows and buffalo. In many parts of India it has been customary for cattle to roam free, or with little supervision, often behaving as scavengers. The dairy cooperatives have provided assured markets and through village-level artificial insemination have increased the quality of the animals. There is now an incentive to provide better feed, and to control diets through stall feeding of cut grasses, tree fodder, crop residues and grain. In India much of the manure is used for fuel, but elsewhere, in East Africa for example, the manure is being used on the farm crops.

Finally, in this list of nutrient-replacing techniques of especial value to the lower-potential areas, is another long-standing practice: composting. Typically, composts consist of mixtures of animal manures, green material and household wastes that are heaped together in such a way as to encourage anaerobic decomposition. The heat generated in the compost heap destroys pathogens and seeds, and the roots of weeds. After the materials are sufficiently decomposed the compost is applied, either in heaps around individual plants or worked into the soil. Composts are especially valuable in tropical climates since the high levels of organic matter help to store nutrients and protect them against leaching. The soil is made more friable and easier to plough and maintains the moisture better.

Although composts are easier to make and apply in kitchen or home gardens, they can be used on a larger scale. The Wafipa people of Tanzania have long practised a system of mound cultivation based on composts [49]. They begin by cutting grass and bushes, which are burnt in small piles. Cucurbits are planted in the ash. Then, in between the cucurbits, mounds of about 90 cm high and 30 cm apart are constructed by piling up turves of grass, the grass facing inwards. Legumes, such as beans and cowpeas, are planted in the mounds, which are left for the grass to rot. When the legumes have been harvested and the rains begin, the mounds are broken down and their contents spread over the land in preparation for sowing with millet or maize. After a couple of years of cultivation the land is left fallow for 4–10 years. The soils are poor, with little clay or natural organic matter and low water-holding capacity.

In an even harsher environment, in the Matengo highlands of southern Tanzania, farmers construct pits on the steep hillsides [50]. The grass is cut and laid in strips, in a grid-like fashion surrounding the pits. Soil from the pits covers the strips and is sown with beans. Weeds and crop residues are placed in the pits to form compost and the pits are sown with maize. Composting in these situations greatly improves their productivity but the process is very labour-demanding. In the Matengo highlands there is no alternative, but the Wafipa people are moving towards use of the plough and inorganic fertilizer, despite its high cost. One answer may be to make composting more efficient and available commercially at prices which are lower than inorganic fertilizer. The Manor House Agricultural Centre in Kenya, which trains farmers in sustainable agricultural practices, has been very successful in promoting the commercial use of composts through its cooperatives. The Pondeni Farmers' Cooperative, for example, consists of a group of farmers who among other activities make and sell compost. It is sieved, mixed with bone meal to provide much-needed phosphorus, and packed in 90 kg bags which sell for $20 each [51].

All these alternatives to inorganic fertilizers have distinct advantages. They are available, or can be created, on or near the farm, and are generated from natural resources. Thus they tend to be relatively inexpensive. They can significantly increase yields, particularly on poor soils, and in some instances will perform as well as or better than inorganic fertilizers. In nearly all situations they are good partial replacements, although it needs to be remembered they may not be less polluting. Nitrates are liable to leaching whether they have an inorganic or organic origin. (In the developed countries some of the worst nitrate-pollution problems arise from the excreta of intensively housed livestock.) However, the main disadvantage of organic fertilizers is their high labour demand. It explains why they have fallen from favour in the developed countries. But in many parts of the developing countries, where, as we have seen, the priority is as much to increase employment and incomes as it is to produce more food, this can be an advantage.

Often arguments about fertilizer use are strongly polarized. On the one hand, many agronomists claim the only way to increase yields is to use large quantities of inorganic fertilizers, and that promulgating the use of organic sources of nutrients will condemn many poor farmers to continuing low yields. Opponents of this view regard inorganic fertilizers as positively harmful and liable to trap farmers into high-cost production. There is some truth in both arguments. Exclusive use of inorganic fertilizers is associated with long-term yield declines; yet where labour is a constraint and where hectarages are large, organic fertilization is insufficient to produce high yields. As in pest management, the approach for the future lies in integration, assessing each situation in agronomic, ecological and socio-economic terms, and then determining an appropriate mix of sources of inputs. This concept – of Integrated Plant Nutrition Systems [52] – is still in its infancy and, like Integrated Pest Management in its early days, is reliant on an expert, top-down approach. But as some NGO programmes have shown, it is amenable to a more participatory approach which will lead not only to greater efficiency but to more sustainability.

Notes

[1] H. B. Ash, 1941, *Lucius Junius Moderatus Columella on Agriculture*, Vols. 1–3, Cambridge, Mass., Harvard University Press (Loeb Classical Library)/London, Heinemann, II, I. 6–13

[2] G. R. Conway and J. N. Pretty, 1991, *Unwelcome Harvest: Agriculture and Pollution*, London, Earthscan (after N. Brady, 1984, *The Nature and Property of Soils*, New York, NY, Macmillan)

[3] D. O. Mitchell and M. D. Ingco, 1993, *The World Food Outlook*, Washington DC, World Bank

[4] FAO/IFA/IFDC, 1992, *Fertilizer Use by Crop*, Rome (Italy), Food and Agriculture Organization (Doc. ESS/Misc./1992/3)

[5] J. L. Jollans, 1985, *Fertilisers in UK Farming*, Reading (UK), Centre for Agricultural Strategy, University of Reading (CAS Report No. 9)

[6] P. J. Stangel, 1979, 'Nitrogen requirement and adequacy of supply for rice production', in IRRI, *Nitrogen and Rice*, Los Banos (Philippines), International Rice Research Institute; N. Roy and S. Seetharaman, 1977, *Wheat* (2nd edn), New Delhi (India), Fertilizer Association of India

[7] FAO, 1989, *Fertilizers and Food Production: Summary Review of Trial and Demonstration Results, 1961–1986*, Rome (Italy), Food and Agriculture Organization (FAO Fertilizer Programme)

[8] N. Alexandratos (ed.), 1995, *World Agriculture: Towards 2010. An FAO Study*, Chichester (UK), Wiley & Sons

[9] R. Prasad and S. K. De Datta, 1979, 'Increasing nitrogen fertilizer efficiency in wetland rice', in IRRI, op. cit.

[10] P. A. Sánchez, 1976, *Properties and Management of Soils in the Tropics*, New York, NY, Wiley & Sons

[11] F. G. Viets, R. P. Humbert and C. E. Nelson, 1967, 'Fertilizers in relation to irrigation', in R. M. Hagan, H. R. Haise and R. W. Edminster, *Irrigation of Agricultural Lands*, Madison, Wis., American Society of Agronomy

[12] S. K. De Datta, 1986, 'Improving nitrogen fertilizer efficiency in lowland rice in tropical Asia', *Fertilizer Research*, 9, 171–86

[13] A. Alexander, 1993, 'Modern trends in special fertilisation practices', *World Agriculture*, 35–8

[14] D. S. Mikkelson, S. K. De Datta and W. N. Obcemea, 1978, 'Ammonia losses from flooded rice soils', *Soil Science Society of America Journal*, 42, 725–30; C. W. Lindau, R. D. Bollich, A. R. DeLaune, A. R. Mosier and K. F. Bronson, 1993, 'Methane mitigation in flooded Louisiana rice fields', *Biology and Fertility of Soils*, 15, 174–8

[15] R. L. Sass, Y. B. Fisher, F. T. Wang, F. T. Turner and M. F. Jud, 1992, 'Methane emission from rice fields: the effect of flood water management', *Global Biogeochemical Cycles*, 6, 249–62

[16] Intergovernmental Panel on Climate Change, 1996, 'Agricultural options for mitigation of greenhouse gas emissions', in *Climate Change 1995: Impacts, Adaptations and Mitigations of Climate Change: Scientific-technical Analyses*, Cambridge (UK), Cambridge University Press, pp. 745–72

[17] R. A. Leng, 1991, *Improving Ruminant Production and Reducing Methane Emissions from Ruminants by Strategic Supplementation*, Washington DC, Office of Air and Radiation (USEPA Report 400/1-91/004)

[18] T. R. Preston and R. A. Leng, 1994, 'Agricultural technology transfer: Perspectives and case studies involving livestock', in J. R. Anderson (ed.), *Agricultural Technology: Policy Issues for the International Community*, Wallingford (UK), CAB International, pp. 267–86

[19] ibid.

[20] Intergovernmental Panel on Climate Change, op. cit.

[21] De Datta, 1986, op. cit.

[22] D. T. O'Brien, M. Sudjadi, J. Sri Adiningsih and Irawan, 1987, 'Economic evaluation of deep placed urea for rice in farmers' fields: a plot area approach, Ngawi, East Java, Indonesia', in IRRI, *Efficiency of Nitrogen Fertilizers for Rice*, Los Banos (Philippines), International Rice Research Institute

[23] IRRI, 1989, *IRRI Toward 2000 and Beyond*, Los Banos (Philippines), International Rice Research Institute

[24] IITA, 1992, *Sustainable Food Production in Sub-Saharan Africa, 1. IITA's Contribution*, Ibadan (Nigeria), International Institute of Tropical Agriculture

[25] D. J. Greenland, 1995, 'Contributions to agricultural productivity and sustainability from research on shifting cultivation, 1960 to present' (unpublished mimeo.); D. M. Robinson and S. J. McKean, 1992, *Shifting Cultivation and Alternatives: An Annotated Bibliography, 1972–1989*, Wallington (UK), CAB International

[26] P. Sánchez, 1994, 'Alternatives to slash and burn: a pragmatic approach for mitigating tropical deforestation', in Anderson op. cit., pp. 451–79

[27] Greenland, op. cit.; C. J. M. G. Pieri, 1992, *Fertility of Soils: A Future for Farming in West Africa*, Berlin (Germany), Springer-Verlag

[28] Conway and Pretty, op. cit.; J. N. Pretty, 1995, *Regenerating Agriculture: Policies and Practice for Sustainability and Self-reliance*, London, Earthscan

[29] Conway and Pretty, op. cit.; S. S. Kohle and B. N. Mitra, 1987, 'Effects of Azolla as an organic source of nitrogen in rice–wheat cropping systems', *Journal of Agronomy and Crop Science*, 159, 212–15

[30] Sánchez, op. cit.

[31] G. O. San Valentin, 1991, 'Utilization of Azolla in rice-based farming systems in the Philippines', in *Asian Experiences in Integrated Plant Nutrition: Report of the Expert Consultation of the Asia Network of Bio and Organic Fertilizers*, Bangkok (Thailand), Regional Office for Asia and the Pacific, Food and Agriculture Organization

[32] P. K. Agarwal and D. P. Garrity, 1987, 'Intercropping of legumes to contribute nitrogen in low-input upland rice-based cropping systems', *International Symposium on Nutrient Management for Food Crop Production in Tropical Farming Systems, Malang, Indonesia*

[33] San Valentin, op. cit.

[34] D. G. Bares, G. Heichel and C. Sheaffer, 1986, 'Nitro alfalfa may foster new cropping system', *News*, 20 November, St Paul, Minn., Minnesota Extension Service

[35] C. F. Yamoah, A. A. Agboola and G. F. Wilson, 1986, 'Nutrient contribution and maize performance in alley cropping systems', *Agroforestry Systems*, 4, 247–54

[36] Ash, op. cit., I, xxiii. 35

[37] Ash, op. cit., II, x. 1

[38] F. Augstburger, 1983, 'Agronomic and environmental potential of manure in the Bolivian valleys and highlands', *Agricultural Ecosystems and Environment*, 10, 335–46

[39] A. Young, 1989, *Agroforestry for Soil Conservation*, Wallingford (UK), CAB International

[40] R. Bunch, 1990, *Low Input Soil Restoration in Honduras: The Cantarras Farmer-to-Farmer Extension Programme*, London, International Institute for Environment and Development (Sustainable Agriculture Programme Gatekeeper Series SA23)

[41] I. A. Craig, 1987, *Pre-rice-crop Green Manuring: a Technology for Soil Improvement*

under Rainfed Conditions in Northeast Thailand, That Phra, Khon Kaen (Thailand), Northeast Rainfed Agricultural Development Project, NEROA

[42] J. Kotschi, A. Water-Bayer, R. Adelheim and U. Hoesle, 1988, *Ecofarming in Agricultural Development*, Eschborn (Germany), GTZ

[43] Agarwal and Garrity, op. cit.

[44] C. Johansen, 1993, 'Two legumes unbind phosphate', *International Ag-Sieve*, 5, 1–3

[45] Conway and Pretty, op. cit.; Pretty, op. cit.

[46] I. Scoones and C. Toulmin, 1993, 'Socio-economic dimensions of nutrient cycling in agropastoral systems in dryland Africa', paper for *ILCA Nutrient Cycling Conference*, Addis Ababa, International Livestock Centre for Africa; R. McCown, G. Haaland and C. de Haan, 1979, 'The interaction between cultivation and livestock production', *Ecological Studies*, 34, 297–332

[47] M. K. Rerkasem and M. B. Rerkasem, 1984, *Organic Manures in Intensive Cropping Systems*, Chiang Mai (Thailand), Multiple Cropping Project, Faculty of Agriculture, University of Chiang Mai

[48] G. C. Wilken, 1987, *Good Farmers: Traditional Agricultural Resource Management in Mexico and Central America*, Berkeley, Calif., University of California Press

[49] A. C. Mbegu, 1996, 'Making the most of compost: a look at *Wafipa* mounds in Tanzania', in C. Reij, L. Scoones and C. Toulmin (eds.), *Sustaining the Soil: Indigenous Soil and Water Conservation in Africa*, London, Earthscan, pp. 134–8

[50] A. E. M. Temu and S. Bisanda, 1996, 'Pit cultivation in the Matengo highlands of Tanzania', in Reij *et al.*, op. cit., pp. 145–50

[51] Kisian'gani quoted in Pretty, op. cit.

[52] FAO, 1991, *Issues and Perspectives in Sustainable Agriculture and Rural Development*, 's-Hertogenbosch (Netherlands), Conference on Agriculture and the Environment, 15–19 April 1991 (Main document, FAO Newsletter)

13 Managing Soil and Water

In Africa . . . many traditional conservation practices continue to be maintained and expanded, whereas modern soil and water conservation facilities are often poorly constructed and not adequately maintained.

– Jan Pronk, Minister for Development Cooperation, The Netherlands [1]

It is relatively easy to sow a seed in a pot of well-structured, organic soil, place it in a greenhouse, protected from pests and pathogens, water and fertilize the growing plant as and when necessary, and be rewarded with a phenomenal crop. Needless to say, conditions on a farm, particularly in the less-well-favoured parts of the world are, far from this 'ideal' environment. Individual farmers can do much to improve their situation – by buying high-quality seed, applying inorganic fertilizer or supplying nutrients from local resources, hand-weeding the crop and adopting an integrated approach to pest and pathogen control. But the biggest challenge lies in achieving a better soil and water environment for their crops.

Technologies that will deliver more efficient and sustainable soil and water management are available. But they are costly, sometimes extremely so, and tend to work only when they are installed and operated in accordance with strict design parameters. Unlike the maintenance of fertility or the control of pests, where farmers can often achieve a great deal acting on their own, sustainable soil and water management usually requires farmers, and indeed whole communities, to act in concert with one another. In some ways it presents the most difficult challenge of the Doubly Green Revolution. As recent history has shown, attempts to improve the soil and water environment for developing-country farmers have often failed, sometimes catastrophically.

Land can become degraded in a number of ways, through:

242

– Water erosion, the principal cause of degradation, accounting for about two-thirds of the total;

– Wind erosion, important in dryland areas, where it is responsible for much of the 'desertification', accounting for another quarter;

– Physical degradation: crusting, compaction, sealing, de-vegetation, excessive tillage, impeded drainage, waterlogging, reduced infiltration and waterholding capacity; and

– Chemical degradation: salinization, alkalinization, acidification, nutrient leaching and depletion, removal of organic matter, burning of vegetative residues, agrochemicals and industrial pollutants.

These categories are clearly interconnected and feed on each other. For example, excessive tillage results in greater erosion and a loss of nutrients.

Identifying land degradation on an individual farm is relatively straight-forward. The topsoil may be eroded through wind or water erosion, or degraded by waterlogging or the accumulation of salt or other toxins. But there are difficulties in extrapolating from the individual farm or experimental plot to the whole catchment [2]. Often soil-loss estimates do not match the levels of river or lake siltation in the same catchment. In many instances, the soil is simply moved from one part of the catchment to another and is not lost to the system [3]. These problems are compounded when we try to assess the extent of degradation at a national or regional level. One set of regional estimates, arising from the Global Land Assessment of Degradation (GLASOD) exercise, puts total global degradation since the Second World War at some 2 billion hectares, or 22·5 per cent of the world's agricultural, pasture, forest and woodland [4]. On these figures, 80 per cent of the crop land in the developing countries – over 400 million hectares – is degraded. And various estimates put the annual loss of land at 5–10 million hectares per year [5].

These figures, if correct, are very worrying. However, some commen-tators, pointing to the weak basis of these estimates, believe they are largely meaningless and certainly exaggerated [6]. A case in point is the common claim that Sub-Saharan Africa is suffering from widespread 'desertification'. Images of 'deserts on the move' have been imprinted in the minds of policy-makers and the public in general. The United Nations Environment Programme reported that 'desertification threatens 35 per cent of the earth's land surface and 20 per cent of its population' [7]. Yet, as Jeremy Swift of IDS argues, much of the data is highly questionable [8]. In particular it often relies on snapshot assessments, comparing drought with wet years, ignoring the often temporary nature of vegetation change, the capacity of dryland ecological systems to recover and the ability of farmers and pastoralists to adapt to the climatic cycles. What may seem to be a desert one year is a productive tract of land in the next.

Tim Dyson also makes the point that the areas of the world which are claimed to be worst affected are not those where yields are currently declining [9]. Most of the severe degradation is on the red-brown soils lying under the rainforests and savannas of the tropics, whereas much of the global grain harvest comes from the black and brown earths of the temperate zone, which tend to be fairly robust to light and even moderate degrees of degradation [10]. Dyson's conclusion is that 'a significant fraction of the land classified by GLASOD as "degraded" lies in parts of the world that are of only marginal importance for aggregate levels of global food production' [11]. However, in many parts of the world soil degradation is a reality and some of the most severely affected regions of the developing world – the uplands and highlands of Asia and Latin America, the semi-arid areas of Sub-Saharan Africa and the saline and waterlogged soils of South Asia – are precisely those where many of the rural poor and chronically undernourished now live. If we are concerned with their future, I believe we have to identify, understand and address soil degradation as it affects them and their livelihoods.

Although losses due to erosion can be considerable, in excess of 50 tons/ ha per year (Figure 13.1), the severity of the consequences depends not only on the nature of the vegetative cover, but also on the depth of intrinsic fertility of the soil. A great deal of loss can be tolerated on deep temperate soils, resulting in relatively small reductions in yield spread over a long period [12]. But losses in tropical countries can be considerable. For the uplands of the island of Java, the on-site costs in terms of lost agricultural production are estimated at $324 million a year (equivalent to 3 per cent of agricultural GDP). To this must be added the off-site costs resulting from silting of the lowland irrigation systems, reservoirs and harbours, which adds a total of up to another $90 million [13]. Rattan Lal, a leading soil scientist based at IITA, has estimated that erosion has caused yield reductions of 2–40 per cent in Africa, and a total loss of cereal production in 1989 of 3·6 million tons [14]. Other studies have shown declines of more than 50 per cent in upland crop yields in some parts of South-East Asia and the Middle East [15].

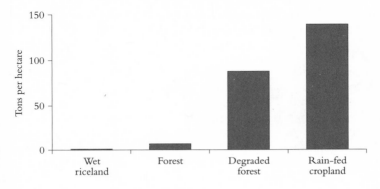

Figure 13.1 Soil losses due to erosion in Java [16]

For many years much time, money and effort has gone into soil-conservation measures in the developing countries [17]. The droughts in southern Africa at the beginning of this century and the experience of the Dust Bowl in the USA in the 1930s stimulated major investments in conservation works by colonial governments, especially in Africa. Between 1929 and 1939 over 7,000 km of bunds were constructed in Southern Rhodesia (now Zimbabwe), and in Nyasaland (now Malawi) nearly 120,000 km were built between 1945 and 1960 [18]. In Lesotho all the uplands were said to be protected by buffer strips by 1960. There was a similar response to the devastating drought that hit Ethiopia and neighbouring countries in the 1980s. Some $20 million was spent annually between 1980 and 1990 on food-for-work programmes in Ethiopia, creating over 200,000 km of terracing. Thousands of hillsides were closed off, steep-slope agriculture was abandoned and 45 million trees were planted [19].

But rarely have these efforts resulted in sustained conservation. Jules Pretty lists, among many, the following recent failures:

— Forty per cent of terraces installed in Ethiopia had been broken within a year;
— Some 120,000 hectares of earth bunds constructed at high cost by machine graders in Burkina Faso in the 1960s have all but disappeared;
— Most of 6,000 hectares of earth bunds constructed in Niger in the 1960s and 1970s are in an advanced state of degradation;

245

– In Sukumuland, Tanzania, almost no evidence remains of a major programme of contour banks, terraces and hedges.

Why is this outcome so common? The techniques of soil and water conservation are well known [20]. Physical structures of varying scale can check the surface flow of water, so reducing water erosion and retaining soil and nutrients. The simplest approach is to throw up earth banks or bunds, 1·5–2 metres wide across the slope and 10–20 metres apart. These are suitable for slopes of 1°–7°. Sometimes they are reinforced with vegetation such as crop stalks or planted with trees to create greater stability. They are not easily damaged and maintenance costs are low. Simple walls may also be constructed along the bunds; after the first heavy rains, fine soils, branches and leaves begin to fill in the walls, making them even more impermeable.

More elaborate physical structures are various forms of terrace, including:

– Diversion terraces, used to intercept overland flow on a hillside and channel it away across a slope to a suitable outlet, appropriate on slopes up to 7°;
– Retention terraces, level terraces used to conserve water by storage on a hillside, on slopes up to 4·5°; and
– Bench terraces, alternating series of shelves and risers used to cultivate steep slopes, the risers often faced with stones or concrete, effective up to 30°.

Sometimes the reason why government–planned and funded conservation fails is that the techniques are inappropriate [21]. Narrow-based terraces in the 1940s were found to fill up with sediments too quickly, were impossible to maintain and began to make erosion worse. John Kerr and N. Sanghi of ICRISAT describe how contour bunds in India were rejected by farmers, even when heavily subsidized [22]. Among various faults are: the bunds leave corners in some fields so that farmers risk losing their land to neighbours; the central watercourses provide benefits to some farmers but damage the land of others; and, if the facilities for dealing with surplus water are inadequate, the bunds readily breach and the water forms gullies. It was not uncommon for entire bunds to be levelled as soon as the project staff had gone to the next village. In Oaxaca, Mexico, contour bunds made erosion so much worse that, eventually, only 5 per cent of the bunded area was cropped.

Part of the problem is the high initial cost of construction and the amount of labour and time needed for maintenance. A cheaper approach is to use vegetation as the means of conservation. The simplest technique is to plant the main crop along the contour, alternating with a protective crop such as a grass or legume. Water flowing down the slope meets with the rows of crops, is slowed down and infiltrates the soil. It is a technique suitable for slopes of 3°–8.5°. Strips of grass will help to filter out particles and nutrients from the water, and over time will build up into terraces. In Indonesian experiments, strips 0·5–1 metre wide of Bahia and signal grass were grown along the contours, alternating with 3–5 metre-wide strips of annual crops [23]. Erosion was reduced by 20 per cent and after four years natural terraces 60 cm high had been formed.

The interplanting of trees and agricultural crops is a technique of great antiquity – the trio of olives, vines and wheat, grown in rows, was the mainstay of classical agriculture in the Roman and Greek empires. But contemporary interest lies in the tropical tree legumes because of their capacity to fix nitrogen. On experiment stations, for example at IITA, fast-growing species such as *Gliricidia* and *Leucaena* (and others listed in Box 12.2) are grown in rows, with 4-metre-wide 'alleys' in between for the annual crops [24]. The trees provide nitrogen, organic matter through their leaf fall and prunings, food for livestock, fuel-wood, and timber, as well as conserving soil and water. The results on experiment stations and demonstration farms are often spectacular but, for the most part, adoption has been poor. This is partly because alley cropping has been developed as a package, whereas farmers tend to be more willing to adopt various components, modifying their farms bit by bit [25]. It is also not enough for the tree crop to provide a soil- and water-conservation benefit. Farmers usually look for an extra direct income as well [26]. Some of the most effective and sustainable examples of the incorporation of trees into arable farming are the home gardens of the tropics (Chapter 9), and the small intensive agricultural plots which I described from the Kakamega district in Kenya (see Figure 2.2).

Trees are also important in reducing wind erosion. Some 140,000 hectares of coastal fields in southern China are protected by windbreaks and shelter belts of such trees as *Casuarina*, *Acacia*, *Leucaena* and certain species of bamboo [27]. They protect against the typhoons in the rainy season and help mitigate the cold spells in late spring and autumn. Wheat and rice yields are said to be 10–25 per cent higher as a result. In Niger, the neem tree (also a source of pesticides, see Chapter 11) is used in a similar fashion.

One of the main advantages of agroforestry is the provision of vegetative material that can be used as a mulch, a cover for the soil [28]. A particularly useful tree in China is *Paulownia*, a relative of the foxgloves; it is very deep-rooting and fast-growing [29]. When mature it will provide up to 400 kg of young branches in a year and 30 kg of leaves. Mulches protect the soil from erosion, desiccation and excessive heating. They can also help reduce the spread of soil-borne diseases, by preventing the splashing of the lower leaves of the crops which spreads the fungal spores. Straw is a common mulch; in the Chiang Mai valley the second crop after the rice – of cabbages, onions, and other high-value vegetables – is heavily mulched with rice straw [30]. One of the reasons why the farmers in the valley retain the traditional rice varieties, albeit improved, is because of the amount of straw produced. A short-strawed variety would produce a greater rice yield, but they would lose income on the second crop. A good mulch is thus much prized. In Guatemala, the farmers of Quetzaltenango collect leaf litter from the nearby mixed pine–oak forests and apply it at a rate of 20–30 tons/ha [31].

The alternative to a mulch is a cover crop, often established after the main crop or as an intercrop. Legume covers will also add nitrogen. The government extension and research service (EPAGRI) in the southern Brazilian state of Santa Catarina has been very successful at promoting cover crops as part of a programme of soil and water conservation [32]. Some sixty different species are involved, mostly legumes but also oats and turnips. They are grown in the fallow period, then knocked over and cut up by a special animal-drawn tool. Another tool, designed by the farmers themselves, is used to plough a narrow furrow, along the contour, in the resulting mulch. The soils are reportedly darker in colour, moist and full of earthworms. Maize yields have risen from 3 to 5 tons/ha.

EPAGRI has been successful because of its involvement of the farmers at all stages in the process. By 1991 EPAGRI was working with some 38,000 farmers in sixty micro-watersheds. It began as a small project, and then expanded rapidly as confidence in the approach grew among the participants. But EPAGRI has been unusual. Far too often government-funded and implemented programmes have lacked an understanding of local ecological and socio-economic conditions and have been unwilling to involve farmers in a manner that allows them to articulate their needs and adapt conservation measures to their requirements [33]. It is not surprising that so many have failed.

Sustainability also depends on embedding conservation within a broader programme of participatory development. In the AKRSP soil- and water-conservation programme in Gujarat in India, that I described in Chapter 10, the village institutions, as they have developed, have increasingly taken on group operations such as plant protection, the pooling of equipment and the marketing of produce [34]. Banks have begun to advance the institutions with loans. Fewer people now engage in seasonal migration off the farms, there is higher school enrolment and improved health and nutrition standards. Leadership in the villages has moved away from the traditional leaders to those who have been most active in the conservation programmes.

Over the centuries, farmers in the developing countries have evolved a great range of conservation systems, adapted to their local conditions. In a recent review, Chris Reij of the Free University in Amsterdam, Ian Scoones of IDS and Camilla Toulmin of IIED have assembled twenty-five accounts by African scientists and extension workers which describe indigenous conservation techniques [35]. They range from the mounds and pits, that I mentioned in the last chapter, to complex systems of bunds and terraces. As the authors point out, these systems, far from being static, evolve as conditions change and will readily incorporate new techniques as and when they are shown to be suitable. Using a technique known as Rapid Catchment Analysis (based on AEA and PRA), Jules Pretty, Jenny McCracken and their Kenyan colleagues demonstrated that farmers in the Murang'a District of Kenya had a sophisticated knowledge of the various techniques available to them: retention ditches, bench terraces, infiltration pits and two local methods, *fanya juu* and *fanya chini* [36]. They were aware of their various advantages and disadvantages, notably the costs and labour involved, and were able to design an integrated soil- and water-conservation system for their watershed, taking these into account.

Far too often, however, indigenous systems are overlooked in official programmes. When Robert Chambers, Jenny McCracken and I worked in the impoverished Wollo region of Ethiopia, we came across many examples of gully plugs – stone walls constructed across the gullies which trap silt, nutrients and water, located high in the hills, and support rich micro-plots of arable crops and trees [37]. They are, indeed, common in dryland areas in many regions of the world – in India and Pakistan and

Nepal, in Burkina Faso and in Mexico, to name some countries [38]. The environments they create — small, flat, fertile and moist fields — are quite unlike the surrounding countryside and can support high-value cereals and cash crops such as coffee or mango. Although construction of the gully plugs is expensive, they are relatively easy to maintain and sustain an agriculture that is productive and dependable. Parmesh Shah of the Aga Khan Rural Support Programme reports that in Gujarat they provide the most stable component of the household's food supply [39]. Yet they are often ignored in conservation programmes, partly because they are not immediately obvious to the visitor [40]. In Ethiopia in the 1980s the government was engaged in moving people off the hillsides and the gully plugs were in danger of being abandoned.

Most of this chapter has, ostensibly, been devoted to soil conservation but, as is clear from the examples I have cited, soil and water conservation are intimately connected. Water is as important for the productivity of plants as is the provision of a good soil structure and sufficient nutrients. A wheat grain may contain up to 25 per cent water, a potato 80 per cent. For rice, in particular, water is crucial: a gram of grain can require as much as 1,400 grams of water for its production [41]. Water stress during growth results in major yield reductions. And the hazard is greater for the new varieties because of their short stature. The young, transplanted seedlings may die for lack of water in the first few weeks and will drown under excessive flooding. Ideally they need a constant flow of water at a depth of about 2·5 cm. Traditional rain-fed cultivation, which is subject to the vagaries of rainfall in the wet season, rarely can provide such exacting circumstances and high yields require supplemental irrigation in most situations. And for all the new cereals, the potential to mature and produce grain irrespective of the season has placed a high premium on the provision of irrigation water in the dry season, when the potential yields are greatest.

Not surprisingly, given the potential returns, the developing countries began to invest heavily in the late 1960s in large-scale, government-designed and operated irrigation systems. The Upper Pampanga River Project in the Philippines, which covers some 80,000 hectares, was completed in 1975 at a cost of over $100 million. The somewhat larger Muda River Project in Malaysia was finished five years earlier. By 1975 the proportion of riceland growing a second, dry-season crop had risen in the Philippines and Malaysia to 60 per cent and 90 per cent respectively [42].

In South Asia, there was also considerable investment, government and private, in tube wells. For the developing countries as a whole, the amount of irrigated land has nearly doubled, as has the proportion of arable land that is irrigated (Figure 13.2). About one-half of developing country cereal production comes from irrigated land [43].

However, as we saw in Chapter 7, there are signs of a slowing down in the rate of expansion of irrigation. Partly this is because of the increasing costs (Figure 13.3). The Upper Pampanga Project in the Philippines covered a command area of 83,000 hectares at a cost of $1,270 per hectare and the Muda River Project in Malaysia serviced 96,000 hectares at an initial cost of $850 and required further investments of another $2,000 per hectare [44]. But much of the topography that is most readily amenable to large-scale irrigation has already been exploited, and costs in most countries of Asia are now over $4,000 per hectare.

Many of these projects have also incurred considerable environmental and social costs. The siting of reservoirs invariably is a major source of contention. Villages and, sometimes, small towns are inundated and the people have to be resettled; more often than not they end up on poor-quality land, and receive few of the benefits from the irrigation scheme. In Chapter 10 I described the conflict over the building and operation of the Buhi dam in the Philippines. But this story is not an isolated one and there are numerous case studies that detail the consequences of large-scale irrigation

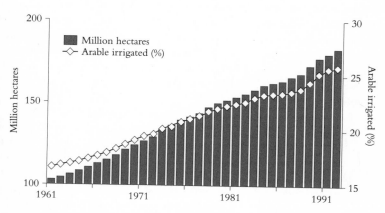

Figure 13.2 Growth in developing-country irrigation

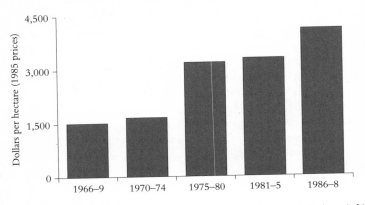

Figure 13.3 Capital costs for construction of new irrigation systems in Indonesia [45]

systems [46]. Often, in addition to the adverse impact on the local population, there are serious losses of forests and wildlife. It is now common
to commission Environmental Impact Assessments. These may run into
several volumes of detailed information, but they are usually conducted
by experts from outside the locality and, as at Buhi, can miss the crucial
issues. By contrast participatory methods, of the kind used at Buhi, can
ensure a much better assessment of the likely effects and a more equitable
and sustainable set of solutions to the problems that will inevitably arise. For
environmental problems do not cease with the completion of construction.
One of the consequences of high rates of erosion is the rapid build-up of
sediments in reservoirs and irrigation canals, so shortening their expected
life. A survey of seven reservoirs in India reveals reductions in the design
lives of 22–94 per cent [47].

As I indicated earlier, a great deal of investment has also gone into the
provision of tube wells. In India the number of wells has increased from
nearly 90,000 in 1950 to over 12 million in 1990 [48]. They have several
advantages over large-scale irrigation systems: they can be relatively easy
and cheap to install and, because they occupy little land, create few
environmental and social problems. But they will only provide a sustainable
supply of irrigation water if the rate of extraction is below that of the rate
of recharge to the underground aquifers from which the water is being

obtained. If this is not the case the water is effectively being mined. In many parts of West and South Asia, over-pumping, encouraged by subsidized electricity, is causing an alarming fall in water levels – a problem that is compounded by falling rates of infiltration to the aquifers resulting from degradation of the upland watersheds. R. S. Narang and M. S. Gill of the Punjab Agricultural University report that over two-thirds of the Punjab the water-table is falling at 20 cm per annum [49]. As a consequence, deeper wells are having to be sunk and the energy costs of extraction are rising.

While in some areas water-tables are falling, in others they are rising, and creating serious problems of waterlogging and salinization. The common cause is a combination of excessive use of water and poor drainage. Cheap, subsidized electricity encourages profligate water use while growing capital costs have resulted in savings in drainage investment, sometimes eliminating drainage systems altogether from new projects. Salinization is usually due to rising water-tables coupled with high rates of evaporation that bring toxic salts to the surface. In some coastal areas, particularly acute salinization is created as a result of over-extraction which leads to salt-water intrusions. The extent of waterlogging and salinization in the developing countries is now considerable. Half the irrigated land in Syria is said to be affected, 30 per cent in Egypt and 15 per cent in Iran [50]. In India the total is 5–13 million hectares or 10–30 per cent [51]. On a worldwide basis, estimates suggest up to 1·5 million hectares are lost annually out of production (equivalent to well over half the newly irrigated land added each year) and 10–15 per cent of irrigated land is to some extent degraded through waterlogging and salinization [52]. Technical solutions are available [53]. The water-table can be lowered below the root zone and the salts flushed away to newly constructed subsurface drainage systems. Costs in India are of the order of $325–$500/ha. But sustainable solutions, as in other areas, depend on local community involvement [54].

The root causes of these economic, social and environmental problems are bad irrigation design and poor maintenance and management. Irrigation, if it is to be effective, has to be reliable, otherwise much or all of the potential benefit will be lost. Reliability depends, in turn, on an efficient and responsive organization, and the question here is whether this can best be supplied under central government or local control. Many of the most ancient of irrigation systems, for example in Mesopotamia, in Sri Lanka and in China, relied on central government control. But equally there are ancient irrigation systems, such as the *subaks* in Bali, that are much smaller – about 200 hectares – and are built and operated by the local community. Such systems tend to be highly responsive, the supply and regulation of the water being an integral part of the traditional practices of resource management in the community. Often the answer lies in creating a good partnership between government and local communities. In the valleys of northern Thailand there is a history going back at least 700 years of community-maintained dams constructed of stone and wood, linked to irrigation systems governed by representative bodies known as *muang*. When in the 1960s and 1970s the government constructed large-scale diversion systems designed to provide year-round irrigation, in many places it grafted them on to the local systems. During our Agroecosystem Analysis of the Chiang Mai Valley (as I recounted in Chapter 10) we discovered that triple-cropping was only being practised in the areas of these joint systems, since it was under these conditions that the water supply was reliable enough to risk planting the high-value third-season crops [55].

Equity, in the sharing of water and other benefits, is also a central and contentious issue in irrigation. Corruption in irrigation administration is widespread, and direct action or bribery is a common means whereby farmers seek to redress perceived imbalances or to gain unplanned or illegal shares. During water shortages in the Minipe Scheme in Sri Lanka farmers have been known to block the channels and divert water to their fields [56]. Sometimes open force may be used. In South India, farmers at the 'tail end' of the irrigation scheme, who are usually those least likely to gain a fair share, have been known to 'hire a jeep and budget for its costs out of their common fund raised for such purposes, and then for farmers to patrol the main canal lowering sluice gates and threatening violence. Occasionally a whole lorryload of farmers will go, brandishing sticks in a demonstration of force' [57].

Bribes are paid to the irrigation engineers, to ensure more reliable supplies, particularly by farmers at the tail ends of the system, and also by farmers wishing to grow unauthorized crops. The bribes may be in the form of an annual flat rate or individual payments. There may also be gifts of grain at harvest, mostly as rewards to the local field staff. In South India the average costs of the bribes in the late 1970s were 4–10 rupees per hectare over two seasons [58]. This is low compared with the 360 rupees net profit to be gained from a rice crop, but the costs can be much higher for tail-enders and are likely to bear heavily on poorer farmers.

Irrigation engineers may also receive 'kickbacks' from contractors under-taking the maintenance of the systems. Niranjan Pant described the practice in the late 1970s on the Kosi Project in Bihar, northern India, where the contractor 'has to spend about 30 per cent of his bill on overseers and engineers, about 10 per cent on office staff, about 10 to 20 per cent for his own profit and thus only 40 to 50 per cent is spent on the actual work' [59]. Both engineers and contractors may make money by colluding in substand-ard work – so-called 'savings on the ground'. Poor-quality cement is used, for example, and the savings are shared. So profitable (Figure 13.4) are pos-itions in the irrigation bureaucracies in many parts of India that the prospec-tive engineers will pay large sums of money in order to be appointed. The posts are usually in the gift of local politicians and, in effect, are auctioned. Robert Wade of IDS, in a detailed study of South India, has revealed payments of three or more times the annual salary to acquire a secure two years in a post. For senior engineers the payment may be as much as forty times the annual salary. The consequences of corruption are not only the higher direct costs incurred by farmers but the negative effects on agricultural production of poorly maintained and inefficiently managed irrigation systems.

Corruption can be minimized by an institutionalized system of com-munity control. One answer, in large irrigation schemes, may lie in delivering water to holding reservoirs before local distribution and placing

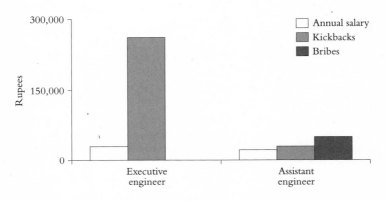

Figure 13.4 Corruption in irrigation in South India [60]

the control over the reservoirs in the hands of local communities. For smaller systems, it becomes easier to institute effective community control, which can extend to design and contract management. An example, still at the experimental stage, is provided by a programme to rehabilitate small tank systems in the south Indian state of Tamil Nadu, where rainfall is less than 850 mm per year and is erratic [61]. The tanks are small, natural, low-lying areas which are dammed so as to catch and store the monsoon rains. Subsequently, the water is used by the villagers to irrigate crop fields. The maintenance of the tanks and the irrigation canals has been the responsibility of government authorities but, for a variety of reasons, the systems have progressively fallen into disrepair. A current aid project is attempting their rehabilitation by hiring contractors who work to a blue-print; not surprisingly this produces inappropriate and excessively costly solutions. In the experiment, several villages have been given grants by the District Rural Development Agency and encouraged, with the technical assistance from an NGO, PRADAN and the Centre for Water Resources at Anna University, to form water-users' associations to design, plan and manage the rehabilitation themselves. In one village, Panchanthangi Patti, the villagers have contributed 25 per cent of the costs in terms of labour, materials and money. They have determined the priorities and identified the work needed – including strengthening the earthen bund, partially desilting the tank and feeder channels, building check dams across the channels to prevent silting and planting trees on the foreshore to prevent encroachment. So far the results are very encouraging. The villagers are showing a high degree of competence and inventiveness, and the outcome is systems which the villagers feel they own and to which they are committed.

The general lesson from the experiences of the past thirty years is that small, community-managed and designed irrigation systems are more likely to deliver sustainable water supplies. In many parts of the world there is, in any case, no other option. Irrigation in Sub-Saharan Africa has grown steadily since the 1960s, but still only amounts to some 6 million hectares – compared with India's 45 million hectares – less than half of 1 per cent of the arable land. Over much of the continent, the environmental conditions are not suitable for large-scale irrigation systems, and the future lies in small-scale systems, like those in Tamil Nadu, and in the drier regions in ingenious systems of water conservation and harvesting based on a micro-catchment approach [62].

Water harvesting from short slopes is relatively straightforward and cheap and can be highly efficient because of the distances involved. In Burkina Faso and Mali various government and NGO projects are helping farmers develop improved water harvesting utilizing the traditional *zaï* system of agriculture [63]. *Zaï*, literally translated as water pockets, are small pits, about 20 cm in diameter and 10 cm deep, placed in lines across the field. In Burkina Faso, under the Djenné Agricultural Systems Project, the *zaï* have been made larger and dug in staggered rows, following the contour lines but perpendicular to the slope, and the excavated earth is used to make small half-moon bunds downslope from each of the pits. On steep slopes they are combined with grass lines and the planting of trees. A similar programme in central Mali, the Project Agro-Forestier begun by Oxfam in 1979, combines the *zaï* with contour bunds constructed of rocks. New techniques of composting have been introduced and a cheap water-tube level has helped in better siting of the bunds. In both programmes yields have greatly increased and become more assured, crops in the *zaï* surviving up to two weeks of drought. And because the systems are being communally designed and managed, they are proving sustainable.

Water harvesting from long slopes requires semi-permeable, stone bunds along the contours, which will slow water run-off and encourage infiltration [64]. In the desert margins of West Asia and North Africa, harvesting targets the periodic flash floods through systems of barriers across the *wadi* floors. Under the Roman Empire, the combination of Roman engineering and local knowledge produced elaborate systems of water harvesting, producing large quantities of wheat and olives, in areas which are, today, largely desert. A modern equivalent in the Central Plateau of Burkina Faso consists of low semi-permeable dams that concentrate and redirect water flows. Natural terraces are formed as the sediment is deposited, on which yields of sorghum increase two- to threefold [65]. The redirection of water is also a feature of the ancient Chinese 'warping systems' [66]. Storm- and flood-waters are diverted around a series of obstacles with the aim of concentrating both water and nutrients. A system covering over 2,000 hectares in Shanxi province provides water in the dry season and helps build up high levels of nutrients and organic matter. Yields of maize, millet and wheat are increased two- to fourfold.

Such systems are demanding of labour, which makes them particularly suitable for lower-potential lands with high population densities. Costs range from $100 to $1,000 per hectare but the returns are considerable:

35 per cent in the Project Agro-Forestier. In all cases the intimacy of water and soil conservation is critical. Adequate water is a necessary but not a sufficient condition: good crop growth also requires considerable supplies of nutrients and a supportive soil structure. The most successful systems are those where techniques are developed which promote all these requirements in a synergistic fashion.

Notes

[1] C. Reij, I. Scoones and C. Toulmin (eds.), 1996, *Sustaining the Soil: Indigenous Soil and Water Conservation in Africa*, London, Earthscan

[2] I. Scoones, C. Reij and C. Toulmin, 1996, 'Sustaining the soil: indigenous soil and water conservation in Africa', in Reij *et al.*, op. cit., pp. 1–27; M. Stocking, 1993, *Soil Erosion in Developing Countries: Where Geomorphology Fears to Tread!*, Norwich (UK), University of East Anglia (School of Development Studies Discussion Paper 241)

[3] J. Bojo and D. Cassells, 1995, *Land Degradation and Rehabilitation in Ethiopia: A Reassessment*, Washington DC, World Bank (AFTES Working Paper 17)

[4] L. R. Oldeman, 1992, *Global Extent of Soil Degradation*, Wageningen, International Soil Reference and Information Centre (Biannual Report); L. R. Oldeman, R. T. A. Hakkleing and W. G. Sombroek, 1990, *World Map of the Status of Human-induced Soil Degradation: An Explanatory Note* (rev. edn), Wageningen (Netherlands), International Soil Reference and Information Centre/Nairobi (Kenya), United Nations Environment Programme

[5] S. J. Scherr and S. Yadav, 1996, *Land Degradation in the Developing World: Implications for Food, Agriculture and Environment to 2020*, Washington DC, International Food Policy Research Institute (Food, Agriculture and Environment Discussion Paper 14)

[6] N. Alexandratos (ed.), 1995, *World Agriculture: Towards 2010. An FAO Study*, Chichester (UK), Wiley & Sons; T. Dyson, 1996, *Population and Food: Global Trends and Future Prospects*, London, Routledge; Scoones *et al.*, op. cit.; M. Leach and R. Mearns (eds.), 1996, *The Lie of the Land, Challenging Received Wisdom on the African Environment*, Oxford, James Currey/Portsmouth, NH, Heinemann

[7] UNEP, 1984, *General Assessment of Progress in the Implementation of the Plan of Action to Combat Desertification 1978–1984: Report of the Executive Director*, Nairobi (Kenya), Governing Council, Twelfth Session, United Nations Environment Programme (UNEP/GC. 12/9)

[8] J. Swift, 1996, 'Desertification: narratives, winners and losers', in Leach and Mearns, op. cit., pp. 73–90

[9] Dyson, op. cit.

[10] P. Crosson and J. R. Anderson, 1992, *Resources and Global Food Prospects*, Washington DC, World Bank (World Technical Paper 194)

[11] Dyson, op. cit.

[12] P. Crosson, 1992, 'United States agriculture and the environment: perspectives on the next twenty years', Washington DC, Resources for the Future (mimeo.); D. O. Mitchell and M. Ingco, 1993, *The World Food Outlook*, Washington DC, World Bank

[13] W. B. Magrath and P. Arens, 1987, *The Costs of Soil Erosion on Java — A Natural Resource Accounting Approach*, Washington DC, World Resources Institute

[14] R. Lal, 1995, 'Erosion–crop productivity relationships for soils of Africa', *American Journal of Soil Science Society*, 59, 661–7

[15] S. N. Yadav and S. J. Scherr, 1995, 'Land degradation in the developing world: is it a threat for food production in the year 2020?', paper presented at the workshop on Land Degradation in the Developing World: Implications for Food, Agriculture and the Environment to the Year 2020, Annapolis, Md, 4–6 April 1995

[16] Magrath and Arens, op. cit.

[17] J. N. Pretty, 1995, *Regenerating Agriculture: Policies and Practice for Sustainability and Self-reliance*, London, Earthscan; Scoones *et al.*, op. cit.

[18] J. R. Whitlow, 1988, 'Soil conservation history in Zimbabwe', *Journal of Soil and Water Conservation*, 43, 299–303; M. Stocking, 1985, 'Soil conservation policy in colonial Africa', *Agricultural History*, 59, 148–61

[19] IUCN, 1990, *Ethiopian Natural Resources Conservation Strategy*, Gland (Switzerland), International Union for the Conservation of Nature and Natural Resources

[20] G. R. Conway and J. N. Pretty, 1991, *Unwelcome Harvest: Agriculture and Pollution*, London, Earthscan

[21] Pretty, op. cit.

[22] J. Kerr and N. K. Sanghi, 1992, *Soil and Water Conservation in India's Semi Arid Tropics*, London, Sustainable Agriculture Programme, International Institute for Environment and Development (Gatekeeper Series SA34)

[23] S. Abujamin *et al.*, 1985, *Contour Grass Strips as a Low-Cost Conservation Practice*, Taiwan, Food and Fertilizer Technology Center (Extension Bulletin No. 225)

[24] IITA, 1992, *Sustainable Food Production in Sub-Saharan Africa, 1. IITA's Contribution*, Ibadan (Nigeria), International Institute of Tropical Agriculture

[25] J. Palmer, 1992, 'The sloping agricultural land technology', in W. Hiemstra, C. Reijntjes and E. Van Der Werf (eds.), *Let Farmers Judge*, London, Intermediate Technology, pp. 151–64

[26] D. J. Greenland, 1995, 'Contributions to agricultural productivity and sustainability from research on shifting cultivation, 1960 to present' (unpub. mimeo.)

[27] S. Luo and C. R. Han, 1990, 'Ecological agriculture in China', in National Research Council, *Sustainable Agriculture and the Environment in the Humid Tropics*, Washington DC, National Academy Press

[28] IITA, op. cit.

[29] Z. Zhaohua, 1988, 'A new farming system-crop/Paulownia intercropping', in *Multipurpose Tree Species for Small Farm Development*, Little Rock, Ark., IDRC/ Winrock

[30] P. Gypmantasiri, A. Wiboonpongse, B. Rerkasem, I. Craig, K. Rerkasem, L. Ganjanapan, M. Titayawan, M. Seetisarn, P. Thani, R. Jaisaard, S. Ongprasert, T. Radnachaless and G. R. Conway, 1980, *An Interdisciplinary Perspective of Cropping Systems in the Chiang Mai Valley: Key Questions for Research*, Chiang Mai (Thailand), Faculty of Agriculture, University of Chiang Mai

[31] G. C. Wilken, 1987, *Good Farmers: Traditional Agricultural Resource Management in Mexico and Central America*, Berkeley, Calif., University of California Press

[32] Pretty, op. cit.; R. Bunch, 1993, *EPAGRI's Work in the State of Santa Catarina, Brazil: Major New Possibilities for Resource-poor Farmers*, Tegucigalpa (Honduras), Cosecha

[33] Scoones *et al.*, op. cit.

[34] Pretty, op. cit.; P. Shah, 1994, 'Participatory Watershed Management in India: the experience of the Aga Khan Rural Support Programme', in I. Scoones and J. Thompson (eds.), *Beyond Farmer First: Rural People's Knowledge, Agricultural Research and Extension Practice* (Kenya), London, Intermediate Technology

[35] Reij, Scoones and Toulmin, op. cit.

[36] J. K. Kiara, M. Seggeros, J. N. Pretty and J. A. McCracken, 1990, *Rapid Catchment Analysis*, Nairobi, Soil and Water Conservation Branch, Ministry of Agriculture

[37] ERCS/IIED, 1988, *Wollo: A Closer Look at Rural Life*, Addis Ababa (Ethiopia), Ethiopian Red Cross Society/London, International Institute for Environment and Development; R. Chambers, 1990, *Microenvironments Unobserved*, London, Sustainable Agriculture Programme, International Institute for Environment and Development (Gatekeeper Series SA 22)

[38] Pretty, op. cit.

[39] P. Shah, G. Bharadwaj and R. Ambastha, 1991, 'Participatory impact monitoring of a soil and water conservation programme by farmers, extension volunteers, and AKRSP', in J. Mascarenhas *et al.* (eds.), *Participatory Rural Appraisal, RRA Notes*, 13, 127–31, London, International Institute for Environment and Development

[40] Kerr and Sanghi, op. cit.

[41] C. R. W. Spedding, 1996, *Agriculture and the Citizen*, London, Chapman & Hall

[42] R. Barker, R. W. Herdt and B. Rose, 1985, *The Rice Economy of Asia*, Washington DC, Resources for the Future

[43] Alexandratos, op. cit.

[44] Barker *et al.*, op. cit.

[45] M. W. Rosengrant and M. Svendsen, 1994, 'Irrigation investment and

management policy for Asia in the 1990s: perspectives for agricultural and irrigation technology policy', in J. R. Anderson (ed.), *Agricultural Technology: Policy Issues for the International Community*, Wallington (UK), CAB International, pp. 402–34

[46] J. Farvar and J. Milton (eds.), 1972, *The Careless Technology: Ecological Aspects of International Development*, Garden City, NY, Natural History Press, Doubleday; B. Amte, 1989, *Cry, the Beloved Narmada*, Chandrapur, Maharashtra (India), Maharogi Sewa Samiti; T. Scudder, 1989, 'Conservation vs. development: river basin projects in Africa', *Environment*, 31, 4–9, 27–32; E. Goldsmith and N. Hilyard, 1984, 1986, *Social and Environmental Effects of Large Dams*, Vols. 1 and 2, Camelford (UK), Wadebridge Ecological Centre

[47] Alexandratos, op. cit.

[48] ibid.

[49] IRRI, 1995, *IRRI 1994–1995: Water: a Looming Crisis*, Los Banos (Philippines), International Rice Research Institute

[50] D. Grigg, 1993, *The World Food Problem* (2nd edn), Oxford, Blackwell

[51] Pretty, op. cit.

[52] Alexandratos, op. cit.

[53] Pretty, op. cit.

[54] K. K. Datta and P. K. Joshi, 1993, 'Problems and prospects of cooperatives in managing degraded lands', *Economic and Political Weekly*, 28, A16–A24

[55] Gypmantasiri *et al.*, op. cit.

[56] P. Wickramasekera, 1981, *Water Management under Channel Irrigation: A Study of the Minipe Settlement in Sri Lanka*, Department of Economics, University of Peredeniya (Sri Lanka) (mimeo.)

[57] Ramamurthy, pers. comm. quoted in Chambers, op. cit.

[58] R. Wade, 1982a, 'The system of administrative and political corruption: canal irrigation in South India', *Journal of Development Studies*, 18, 287–328

[59] N. Pant, 1981, *Some Aspects of Irrigation Administration (A Case Study of Kosi Project)*, Calcutta-6 (India), Naya Prokash

[60] R. Wade, 1982a, op. cit., 1982b, 'Group action in irrigation', *Economic and Political Weekly*, 25 September (Review of Agriculture) (A103–8); 1982c, 'Corruption: where does the money go?' *Economic and Political Weekly*, 17, 40, 1606; 1984, *The Market for Public Office: Why the Indian State is Not Better at Development*, Sussex, Institute for Development Studies (IDS Discussion Paper 194)

[61] Ford Foundation, 1994, 'Saving the village tank', *Bulletin, New Delhi Office*, 1, 3–5

[62] Pretty, op. cit.; C. Reij, 1991, *Indigenous Soil and Water Conservation in Africa*, London, Sustainable Agriculture Programme, International Institute for Environment and Development (Gatekeeper Series SA27); Reij *et al.*, op. cit.

[63] Pretty, op. cit.; P. Gubbels, 1994, 'Farmer-driven research and the Project

Agro-Forestier in Burkina Faso', in Scoones and Thompson, op. cit:; J. Wedum, Y. Doumba, B. Sanogo, Dicko and O. Cissé, 1996, 'Rehabilitating degraded land: *Zaï* in the Djenné Circle of Mali', in Reij *et al.*, op. cit., pp. 62–8

[64] E. Bastian and W. Gräfe, 1989, 'Afforestation with multipurpose trees in *media lunas*: a case study from the Tarija basin, Bolivia', *Agroforestry Systems*, 9, 93–126

[65] I. Scoones, 1991, 'Wetlands in drylands: key resources for agricultural and pastoral production in Africa', *Ambio*, 20, 366–71

[66] Pretty, op. cit.

14   Conserving Natural Resources

> Technology is good only if it is sustainable and it is sustainable only
> if it includes people, because people are part of the environment.
> Participatory management has succeeded because it is based on the
> belief that people are important and must be involved in the solutions
> to problems.
>
> — Ajit Banarjee, former chief forester, West Bengal [1]

Natural resources come in many forms. Some are living resources such as
the natural enemies that control pests and pathogens (Chapter 11) and the
bacteria and other micro-organisms that fix atmospheric nitrogen (Chapter
12); others are physical resources, notably soil and water (the previous
chapter). Added to this list, and discussed in this chapter, is the world's
rich endowment of biodiversity, contained in the wildlife of our rangelands
and forests, and in the fisheries and other aquatic life of the seas and fresh
waters.

Rangelands, forests and fisheries often are examples of common property
resources. According to the conventional wisdom, common ownership is
the major cause of natural-resource degradation. Each user tends to
maximize his or her 'share' of the resource without regard to other users,
or even to his or her own future use. On a communally owned rangeland
each individual has nothing to lose from increasing the numbers of his or
her own livestock on the pasture, even if others are doing the same. In
the aggregate this will eventually lead to collective overstocking, with
resulting degradation of the pasture. The individual may perceive this
consequence but cannot prevent it by unilaterally limiting his or her herd.
Only the community can act and many traditional pastoral livestock systems
have evolved highly organized controls on the use of common-property
land, with sanctions by the community against individual over-
exploitation. Where such traditional controls persist, the sustainability of
the resource is less of a problem. However, growing populations, increasing

competition between different communities, technological change and insensitive government interventions are undermining traditional institutions. As a result, sanctions and other controls have broken down and many long-standing, common-property resources – rangelands and forests in particular – are becoming effectively 'open access', and hence liable to rapid and severe degradation.

Rangelands in the developing countries are commonly characterized by low or erratic precipitation, poor drainage, extreme temperatures, rough topography and other physical limitations which render them unsuitable for cultivation. But they support most of the developing world's population of cattle, sheep and goats (Figure 14.1). They are also an important source of fuel-wood and a variety of other natural resources [2].

In a country such as Pakistan they include the alpine pastures in the northern mountains, temperate and mediterranean grazing lands in the west, and arid or semi-arid ranges in the Indus valley [3]. Variation in available moisture and nutrients in Sub-Saharan Africa produces a similar mosaic of rangelands (Figure 14.2). Some are rich natural grass swards, others are covered with very rough and sparse vegetation. Most have varying extents of shrub and tree cover and grade into wood lots and forests. In total they occupy some 2 billion hectares, roughly equivalent to the amount of forest land in the developing countries (Figure 14.3).

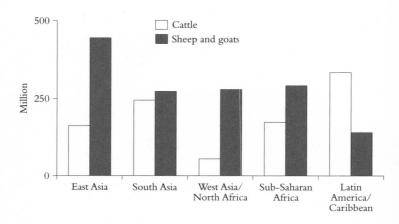

Figure 14.1 Cattle, sheep and goat numbers in the developing world

A great deal of this land is, depending on the purpose to which it is put, degraded. For many years the conventional wisdom has attributed this to overgrazing, but recent research has suggested the concepts of overgrazing and degradation as applied to rangelands in the developing countries have been seriously oversimplified. At the heart of the problem has been the misapplication of management theories developed to maximize beef production on temperate grasslands in the developed countries. In 1979, Graeme Caughley proposed a model of the relationship between animal numbers and vegetation which distinguishes ecological and economic carrying capacity (Figure 14.4). At the far right of the upper curve there is a small population of animals and a large standing crop of plants. As the numbers of animals increase, the amount of vegetation decreases until a

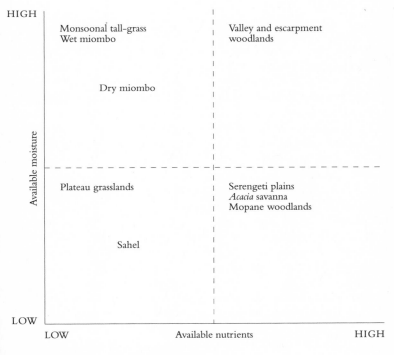

Figure 14.2 Savannas of Sub-Saharan Africa [4]

265

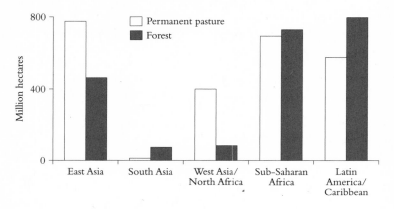

Figure 14.3 Rangeland and forest in the developing world

point is reached – the ecological carrying capacity – where the production of forage equals its consumption and the animal population ceases to grow. At this point there will be many animals but they may not be in good condition and the vegetation will be much less abundant. If the range manager wants denser vegetation or healthier animals, then animals must be hunted, in the case of wildlife, or culled if they are domestic livestock. The lower, offtake curve indicates the various combinations of animals and vegetation that will produce a sustainable meat yield; its maximum is the economic carrying capacity.

For beef production in temperate pastures, being close to the economic carrying capacity is the appropriate goal, but in Sub-Saharan Africa and elsewhere in the developing world the objectives and the circumstances, particularly the climatic conditions, may produce a very different target. If a rangeland, such as the Serengeti savanna in East Africa, is being managed as a wildlife park and a tourist attraction the aim may be to maximize the numbers and diversity of wildlife, the target being, in the words of Roy Behnke and Ian Scoones, a 'camera-carrying capacity', lying somewhere towards the ecological carrying capacity [5]. On the other hand, if the wildlife are being culled for meat, then a much lower population, close to the economic carrying capacity, will be logical. Other aims – to produce high-grade meat, or trophy specimens for hunters, or to preserve special

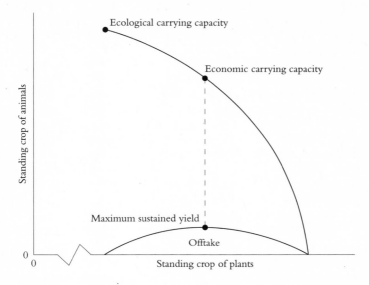

Figure 14.4 The relationship between animals and plants in a grazing system [6]

plant communities – will result in different targets on the curve. Finally, for the pastoralist, the principal user of much of the world's range-- lands, where the aim is to maximize a range of livestock products – milk, blood, traction power and transport – other than meat, it is profitable to maintain a large stocking rate since the offtake does not usually require slaughter.

In these highly different circumstances, 'overgrazing' is relative to the aims of the rangeland managers, as is 'degradation'. There is no objective measure of either overgrazing or degradation. Inevitably, the state of the vegetation will vary considerably, being much poorer at the ecological carrying capacity than at the other extreme, where the range is managed for trophy hunters or high-grade meat. The critical question for sustainability is how reversible are the soil and vegetation conditions, simply by reducing the stocking rate. Crude measures of vegetation are weak guides; changes in the physical and chemical composition of soils and rates of soil erosion are more reliable (see previous chapter), but are compounded by the

267

occurrence, particularly on upland ranges, such as in the relatively young mountains of the Karakoram in northern Pakistan, of high rates of natural, geological erosion.

Many rangelands also exist in regions where the rainfall is highly variable and unreliable (Figure 14.5). In these circumstances there is no permanent target point on Caughley's curve to aim at – the curves move with the rainfall. The state of the vegetation changes dramatically from year to year, often more as a result of the rainfall than the animal stocking density. Stable conditions are rare and carrying capacities difficult, if not impossible, to estimate. In addition to climatic variability, outbreaks of diseases, political upheavals and policy changes add to the disturbances. Pastoral livestock populations thus tend to go through periodic cycles of 'boom and bust'. During the colonial period in countries such as Zimbabwe, tick-borne diseases and devastating epidemics of rinderpest and pleuropneumonia kept the growth of livestock in check. But then disease-eradication campaigns in the 1950s and 1960s, based on dipping the cattle to control the ticks, generated a rapid expansion in numbers, by 2·8 per cent per year in the 1960s, slowing to 1·5 per cent in the 1970s, with periods of local collapse following drought and political upheaval (Box 14.1) [7].

For far too long, African and other developing country rangelands have been subject to policies based on temperate, beef-producing pasture systems. These have led to the encouragement of individual land ownership, reinforced by construction of fencing and other measures to restrict free-range grazing. But such an approach is rarely appropriate. The large numbers of people involved in pastoralism, the diversity of products being sought and the great unpredictability of climatic conditions dictate an approach which is flexible and opportunistic. Pastoralism, with systems of management evolved for these circumstances, needs improvement and support rather than replacement [8].

In a detailed study of pastoral communities in southern Zimbabwe, Ian Scoones, then at Imperial College, London, has shown the remarkable array of adaptive strategies used to cope with uncertainty [9]. The land traditionally exploited by the pastoralists is a complex mosaic of different habitats. In normal years, and when droughts are not too severe, livestock are moved from the top lands to river banks, drainage lines and valley wetlands or bottom lands, known as *dambos*, where moisture levels are high and grassland available. Sometimes animals have to be moved between ecological zones – from the clay soils, where grass production collapses

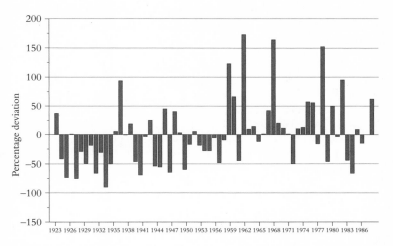

Figure 14.5 Rainfall at Lodwar, Kenya (measured as percentage deviation from the long term mean of 181·01 mm/year) [10]

early in a drought, to sandy soils where production is maintained. In a particularly severe drought, when all local grazing areas are exhausted, livestock may be moved over considerable distances to seek suitable grazing land. Other adaptive strategies include supplementary feeding with hay, cut grass and the branches, leaves and fruits of trees – including the pods of *Acacia* – and slaughter and sale of cattle as droughts set in. Losses during severe droughts can be considerable and the instability of the system is high, but pastoralists, by dint of circumstance, have learnt to be opportunists and can cope, provided government policies enable and do not interfere [11].

In their review of rangeland management in Africa, Roy Behnke and Ian Scoones comment that

International development agencies and African governments have devoted considerable effort to the suppression of pastoral techniques of land and livestock management. These programmes were undertaken in the presumption that pastoralism was inherently unproductive and ecologically destructive and, hence, required radical reform. Current empirical research supports none of these presumptions. [12]

269

Box 14.1 **Fluctuations in cattle populations in southern Zimbabwe over the past 100 years** [13]

pre-1896	High cattle populations, competing with wildlife
1896	Devastation by rinderpest
1896–1945	Recovery of populations, beginning of dipping in 1920s
1945–60	Destocking policy once again reduced populations
1961–75	Abandonment of policies, and high rainfall, growth again
1976–9	Liberation war, abandonment of dipping, populations fall
1980–2	Independence, dipping resumed, populations grow
1983–4	Major drought with large-scale mortality
1985–6	Good rainfall years allow recovery
1987	Mortality from a further drought
1988–90	Recovery
1991–2	Major drought with large-scale mortality
1993	Recovery

Indeed, most studies show pastoral productivity per unit of land area to equal or exceed that of commercial ranching in comparable ecosystems. What is required is a change in attitudes. Government policies can be supportive if they are aimed at creating greater stability and reducing losses. As a first step, governments need to recognize explicitly the central importance to pastoralism of the heterogeneity of rangelands. This would have far-reaching implications, not least for land tenure and the respect for traditional land-use rights. It would result, in times of drought, in the removal of restrictions on movements between areas. And it would justify programmes aimed at protecting and enhancing the key drought resources – the riverbanks, *dambos* and drainage lines – by reseeding degraded areas with improved species. Explicit recognition of the inherent instability of pastoral systems would also encourage governments to support supplementary feeding during severe droughts, through the provision of feed and the establishment of feeding pens, and to facilitate rapid destocking and restocking at the beginning and end of droughts through livestock markets designed for this purpose [14].

Although many forests in the developing countries are managed, like rangelands, as common properties, a high proportion are in state or private ownership. This has not necessarily saved them from degradation. The developing countries possess some 2 billion hectares of forest, that is, land with a minimum of 10 per cent of crown coverage of trees, plus another 1·1 billion of land with some wooded vegetation (the latter is included under rangeland in Figure 14.3) [15]. This represents over 60 per cent of the world's forest and wooded land, but it is not evenly distributed and is being lost at an accelerating rate. Nearly half of the tropical forest lies in South America, much of the rest in Central and West Africa and the islands of South-East Asia. There have been efforts in many countries to replace logged or otherwise destroyed forest with forest plantations, but the rate of replacement is very slow – about 2·6 million hectares per year. The total area of well-established forest plantations is just over 30 million hectares [16].

Much of the loss in Latin America has been due to deliberate government policies of opening up forest land. In the early 1980s, over a million migrants moved into the state of Rondônia in Brazil, clearing an area of land the size of West Virginia. They were offered free land and government support under the Polonoroeste Project, but few of the agricultural practices were sustainable and many of the settlers soon abandoned the land, migrating to the cities or moving on into new areas of forest.

In addition to commercial logging and government settlement schemes, forests can be destroyed through the progressive conversion of forest margins to subsistence agriculture and rough grazing, and as a result of over-cutting for fuel-wood and charcoal production [17]. Yet, as with soil degradation (see the previous chapter), many of the estimates of loss are exaggerated. The 'wood-fuel crisis', like 'desertification', has become part of the conventional wisdom of environmental deterioration. Detailed studies such as those carried out by James Fairhead of the School of Oriental and African Studies in London and Melissa Leach of IDS suggest a more complex picture of change and adaptation. Indeed, in the Kissidougou prefecture of Guniée, where they worked, they demonstrated the expansion of forest resources. Farmers were creating forest islands around their villages and encouraging fallow vegetation to be more woody. The savannas, far from being half empty of forests, were half full [18].

If sustainably managed, the potential income and employment from forest products is considerable (Figure 14.6). Timber, pulp and paper exports earn some $13 billion and an estimated 3 billion people depend on wood for fuel. For the developing countries as a whole, harvested fuel-wood is equivalent to nearly half a billion tons of oil and supplies some 15 per cent of their energy needs, and in the poorest countries this can be as high as 70 per cent [19]. Fuel-wood shortages are now becoming severe. According to FAO, nearly 250 million people suffer from acute shortages of fuel-wood and a further 1·3 billion are living in areas where current demand exceeds the rate of regeneration [20].

Forests are also the source of a wide variety of so-called 'minor or non-timber forest products', such as resins and gums, oil seeds, honey, silk, wild fruits and mushrooms, spices and medicinal plants. These tend not to appear in the official statistics but they frequently provide much-needed income and resources for the poor, and are a major source of employment. Total production in India is estimated to be some 4 million tons, involving 2 million person-days of employment. There is a considerable potential for expansion of production, by up to threefold, and of employment by over twofold (Figure 14.7) [21]. The most important of the non-timber products in South-East Asia is rattan, contributing to a world trade of about $2 billion in value.

Finally, forest lands play a role, much more difficult to quantify, in trapping and storing rainfall, conserving watersheds, preventing erosion, and modifying the climate on both a local and a regional scale. They lock

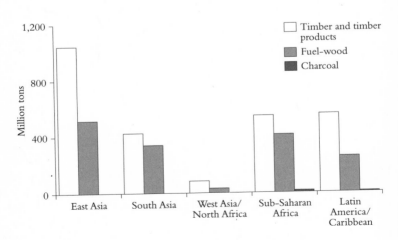

Figure 14.6 Forest production in the developing world

up large quantities of carbon and their destruction for agriculture and other purposes releases carbon dioxide to the atmosphere and hence contributes to global warming. And, since they are also home to a high proportion of the world's fauna and flora, their destruction is one of the prime causes of the increasing loss of global biodiversity. An estimated 15 per cent of the world's plant and animal species could become extinct by 2025, many with potential for agricultural or forest exploitation. This represents not only a loss of useful organisms, for example the predators and parasites that provide natural control of our pests, but more fundamentally a destruction of the world's treasure trove of DNA. As I argued in Chapter 8, genetic engineering holds out the promise of recombining the genetic stock contained in our plants and animals to provide new varieties of crops and livestock. While some recombinations involve moving genes between existing domestic varieties, the greatest potential lies in identifying and isolating useful genes in wild relatives and then transferring them to the domestic stock. Wild rices have already proven to be useful sources of pest

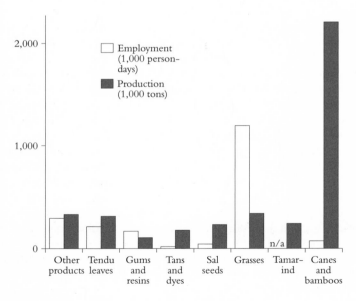

Figure 14.7 Non-wood forest products in India [22]

273

and disease resistance. While the relatives of the domestic grains are usually to be found on rangelands and at the forest margins, for other crops, such as oil palm and cacao, the relatives are usually deep in the primary forest. Currently protected forest areas – including national parks and reserves of various kinds – cover just over a quarter of a billion hectares, little more than 5 per cent of the total land area in the tropics [23].

Forests are thus highly valuable resources if managed on a sustainable basis. Yet the temptation to clear cut, sell and move on is considerable. Great fortunes can be made in this way and it is not surprising that, despite governmental regulations, corruption is an endemic problem. In many parts of the developing world, but particularly in South-East Asia and Latin America, large tracts of land have been granted or leased to wealthy individuals or local companies, often with strong political connections. While the grants or leases usually carry with them the requirement to reforest and manage on a long-term sustainable basis, this is often flouted. Foreign companies have also been guilty of poor practices, although in recent years several of the multinational timber corporations have begun to behave responsibly under pressure from developing and home-country governments and NGOs.

Where governments have exerted direct control over forest lands the situation has not been a great deal better, and in some instances has been worse. In India, a high proportion of state forest land is heavily degraded. Although designated as protected and guarded, bribery and intimidation have allowed local communities unfettered access. Since these communities have no responsibility, either of a traditional or legal nature, for the land and their incursions are illegal, not surprisingly they do not regard themselves as bound to follow sustainable practices.

The answer in such situations lies in giving local communities formal rights and responsibility for management in partnership with government forest agencies. This approach has been very effectively pioneered in the state of West Bengal in India under the title of Joint Forest Management [24]. Hundreds of thousands of hectares of former sal forest (*Shorea robusta*) in the south-west of the state had become denuded of trees and reduced to rough grazing. Increased policing had not solved the problem and in the early 1970s an experiment was conducted at Arabari, by the local chief forester, Ajit Banarjee, in which village communities were offered a deal: if they formed a forest-protection committee to guard and manage the forest they could in return have access to all the minor forest products and receive a 25 per cent share in the final timber products when the forest grew back. This worked well and similar experiments were started in other parts of the state. In the 1980s, the approach was actively supported by the state government, a leftist coalition committed to land reform and other populist programmes. Today, 1,600 rural communities are now responsible for some 80,000 hectares of natural sal forest. Once the animals are removed, the vegetation grows back rapidly, the sal trees springing up from the remnant stumps in the ground and achieving a height of over 20 feet in three years. The overall forest cover in the affected regions of West Bengal is growing rapidly, the aims of the Forest Department are being achieved and the income of the villagers is growing. Women, in particular, are benefiting from a steady flow of income from such products as firewood, oils and seeds, silk and leaves for plate-making. As in other forest lands the so-called minor forest products collectively are of considerable value. The experiment has been sub-sequently replicated in other states of India and has now become a national policy.

Trees, of course, do not grow only in forests and wood lots. Of especial importance in the developing countries are the trees planted on arable land. In the previous two chapters I discussed the value of trees as sources of nutrients and as agents of soil conservation when interplanted with annual crops. Plantation-tree crops, such as rubber, oil palm and coconuts, are important sources of income and, in most conditions, also serve to reduce erosion and to conserve moisture. In Asia they cover some 14 million hectares of land [25]. For the poor, of greatest values are the trees they plant in their intensively cultivated plots and, in particular, in their home gardens. From the air the numerous small plots of the farmers in Kakamega, in western Kenya (described in Chapter 2), appear to be a forest, such is the density of the trees they have planted. As the preference ranking produced by Ethiopian villagers demonstrated (see Chapter 10), rural people have a highly sophisticated understanding of the various properties and uses of trees of many kinds. They will obtain seed or cuttings of desired species and plant them on their land, provided they are assured of ownership and the right to harvest the products. In several Indian states misguided laws aimed at conservation have banned the felling of trees and, as a result, removed the incentive for tree planting [26].

Although some range and forest lands extend across national boundaries, most are subject to national jurisdictions and can, given the will, be managed sustainably through sensitive government policies. This is also true, at least in theory, for the world's marine fisheries, since most of the highly productive fisheries lie within the 200-mile Exclusive Economic Zones that define national fish stocks. However, in practice, marine fisheries present the problem of open-access management on a scale which, despite the availability of legal agreements and technical solutions, is extremely difficult to solve. Not surprisingly, most ocean fish stocks are being rapidly depleted, largely through overfishing.

The global harvest of fish (including crustacea and molluscs) captured in the oceans and inland waters peaked at nearly 89 million tons in 1989 and declined to 85 million tons in 1993 (Figure 14.8). For most of the world's wild fish stocks the harvest is stagnant or declining. Fish farming or aquaculture is rising rapidly, reaching over 16 million tons in 1993 and contributing to a total global fish harvest of over 100 million tons. But the fish harvest per person appears to have stagnated.

Although fish are not an important source of energy in the diet, they provide on average about 19 per cent of worldwide animal protein consumption, and in China and many other parts of the developing world they contribute over half of the animal protein. Fish are also an important source of vitamins, minerals and fatty acids [27]. Of the total global fish harvest, some 100 million tons, about 60 million tons are harvested by the developing countries. This nearly equals the 70 million tons of animal meat they produce from cattle, sheep, pigs and poultry.

Fisheries, like rangelands, can be conceptualized in terms of a range of possible carrying capacities and sustainable yields, depending on the objectives. If preservation is desired, for example of the world's whale stocks, an ecological carrying capacity can be sought; it is also possible to maximize the production of high-quality sport fish, or of small 'industrial' fish. The recent history of the world's marine fisheries has been an accelerating trend towards industrial fishing, harvesting smaller and smaller fish, not for direct human consumption but for feed. As a consequence, the catch of high-value fish has declined. In the Gulf of Thailand the large food fish have been replaced by small, short-lived fish, squid and shrimps [28]. Twenty per cent of world production now consists of small pelagic species used for making fishmeal, which, in turn, is used in pig and poultry production and in salmon and shrimp aquaculture. Two of the most important of these pelagic species are the anchovy and the South American pilchard, which make up the bulk of the catch of Chile and Peru.

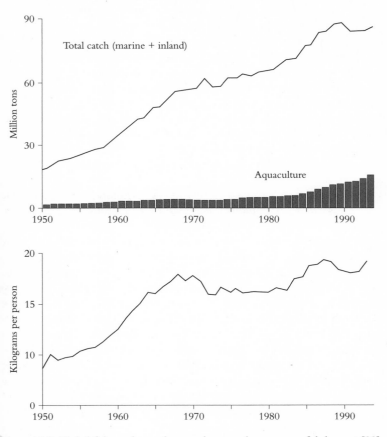

Figure 14.8 Global fish catches and aquaculture, and per person fish harvest [29]

Despite the apparent stability of the oceans, their fish and other populations are as much subject to fluctuation as are rangeland cattle. One of the most productive fisheries in the world, providing 20 per cent of the world's fish landings in the 1960s and 1970s, is generated by upwellings of cold, nutrient-rich waters off the coast of Chile and Peru [30]. The upwellings support rich populations of plankton on which the pelagic

277

anchovies and pilchards feed. Peru's catch of anchovies grew in the 1960s to a total of over 12 million tons, but in 1972 it suddenly collapsed to a mere 2·5 million tons (the fall in production shows up in the worldwide data in Figure 14.8) [31]. For the next ten years the catch only averaged 1·2 million tons. The immediate cause was the arrival of 'El Niño', a phenomenon of the Pacific Ocean which produces persistent warm surface waters, so greatly reducing plankton growth. However, 'El Niño' was probably not the reason why the fall in catch was so large and persistent [32]. In the 1960s, fishery biologists had estimated a maximum sustainable yield of some 9·5 million tons per year and it appears to be the fishing in excess of this level that placed the anchovy stock in jeopardy, making it vulnerable to the climatic change.

Managing fisheries subject to such great instability is, like managing rangelands and forests, a difficult task requiring continuous adaptation and flexible responses to changing circumstances [33]. While it is possible to define clear objectives and identify precise, sustainable targets, they are rarely achievable and it is easy to get into an incremental situation in which steadily increasing returns are suddenly replaced by collapse. Part of the answer relies in creating appropriate policies and regulations, often at international level. A beginning was made in the 1970s with the creation of 200-mile offshore Exclusive Economic Zones, but effective sustainable management has only arisen where governments have got together to agree on targets, and been ready and able to police them. Once targets have been set, it is possible to find effective technical means to achieve them, whether through quotas, size of fishing fleet or types of net. But most important is the need to be flexible and to be cautious, and to learn rapidly. In general it pays, in the long run, to fish at somewhat less than the maximum sustainable yield, so providing a cushion in the face of unexpected events and extreme variability.

Some of the most threatened marine fisheries are those lying inshore that have traditionally been fished by small fishing vessels. The numbers of such vessels has grown, in part because of the landless seeking new opportunities, but they are under considerable threat from competition and environmental degradation. Despite the existence of 200-mile zones, large fishing fleets from the developed countries have often engaged in aggressive fishing. Of twenty-eight recent fishing disputes around the globe nearly one-half involve developing countries (Figure 14.9). But perhaps more insidious and long term in its effect on the sustainability of coastal fishing is the continuing degradation of coastal waters, arising from on-shore developments and river-borne industrial pollutants.

Inland fisheries (lakes and rivers), which are concentrated in Asia and Africa, also appear to have peaked and are showing signs of severe overfishing. Partial compensation for the reduction in marine and inland fisheries has come from the growth in aquaculture, both freshwater and marine, which took off in the 1980s (Figure 14.8). Much of the developing-country growth has occurred in China, partly based on the long-standing tradition of raising herbivorous carp in conjunction with agriculture [34]. But there have also been large increases in shrimp production, where China produces 27 per cent of global output, and mussels, where China produces 38 per cent. Despite its ancient origins, large-scale aquaculture is still focused on only a limited range of species that have yet to benefit from intensive breeding programmes.

In the rice-growing regions of Asia there is a long tradition of raising fish in the wet paddy fields. It can be a highly symbiotic relationship, the fish provided with a nutrient-rich and safe environment and, in return, eating weeds and insect pests and through their excreta increasing the nutrient levels for crop production. Fish will graze on azolla when present and convert it to available nitrogen for the rice plants. In the past, farmers have often simply exploited wild fish populations, constructing small ponds as refuges for the fish in the dry season. Today, high levels of production in ricefields are possible by purchasing good-quality fish stock, constructing nurseries to raise the fry before release, use of supplementary feeding and careful control of stocking rates [35]. The biggest obstacle is pesticide use which in most cases either directly kills the fish or destroys the richness of the habitat on which they depend. A great deal of effort is now going into demonstrating the combined benefits of Integrated Pest Management and fish culture in ricefields. Under a CARE programme in Bangladesh the elimination of pesticides has not only increased fish yields but raised rice yields by some 25 per cent [36].

By contrast with the open sea, where regulation of the fishery is difficult, farmed fisheries avoid problems of over-exploitation. Ownership is usually not in question, stocks can be kept at an optimal size, feed controlled and harvesting timed to gain the maximum return. However, the sustainability of aquaculture is being threatened by inadequate management and by pollution and conflicts over land use, particularly in coastal ecosystems. Clearing of land for shrimp farming has been highly destructive of

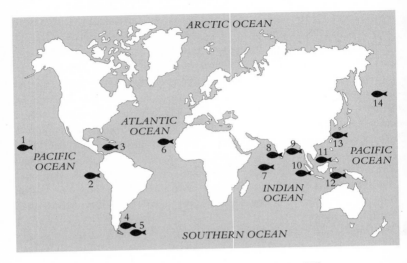

Figure 14.9 Fishery disputes involving developing countries [37]

1. USA threatens China because of failure to implement drift-net ban. 2. Conflicts between commercial and small-scale fishing off Ecuador. 3. Illegal fishing by Japan, Russia, Taiwan, USA and Venezuela in Caribbean. 4. Illegal fishing by Russia and Chile off Argentina. 5. Illegal fishing by Russia and Chile off South Georgia. 6. Illegal fishing by EU vessels off Senegal. 7. Japanese long-liner caught fishing illegally in Indian Ocean. 8. Conflict between industrial and small-scale fishing off Kerala. 9. Pirate attacks on shrimp trawlers in Bay of Bengal. 10. Illegal surimi trawlers off Malaysia. 11. Fishing disputes between China and Vietnam. 12. Taiwan allegedly illegally fishing off Indonesia. 13. Conflicts between Chinese and Taiwanese boats. 14. Poaching by South Korea, China and Taiwan in Bering Sea

mangrove forests – converting over 17 per cent of Thailand's forests in just six years in the late 1980s – and has resulted in rapidly falling water tables [38].

Shrimp farming is especially subject to chronic disease problems. In several Asian countries outbreaks of disease because of poor hygiene and quarantine, coupled with lack of control over water intakes and pond effluents, have contributed to irreversible crashes in production after only two to three years. In Taiwan there has still not been a recovery after the collapse of shrimp production, from a peak of 80,000 tons in 1987 to very limited production in 1991 [39].

Fishery problems highlight the general point that resource degradation often results from overexploitation and conflicts over resource use. In the coastal zones of the developing countries conflicts arise between intensive fisheries, rice production and the natural productivity of mangrove and other swamp forests. Inland, small farmers and large landowners cut down forests to make way for crops and livestock. And worldwide there is growing competition between agriculture and natural resources, on the one hand, and expanding urbanization and industrialization, on the other. Although inappropriate technologies are partly to blame, more fundamentally the causes lie in inappropriate systems for resource management, unresponsive institutions, short-term national and regional policies and a lack of economic mechanisms that will adequately value natural resources in relation to all their potential uses, now and in the future.

Notes

[1] Ford Foundation, 1991, 'Saving the forests: India's experiment in cooperation', *The Ford Foundation Letter*, 22, 1–5, 12–13

[2] L. A. Stoddart, A. D. Smith and T. W. Box, 1975, *Range Management*, New York, NY, McGraw-Hill

[3] N. Mohammed, 1989, *Rangeland Management in Pakistan*, Kathmandu (Nepal), International Centre for Integrated Mountain Development

[4] R. H. Behnke and I. Scoones, 1993, 'Rethinking range ecology: implications for rangeland management in Africa', in R. H. Behnke, I. Scoones and C. Kerven (eds.), *Rangeland Ecology at Disequilibrium: New Models of Natural Variability and Pastoral Adaptation in African Savannas*, London, Overseas Development Institute, pp. 1–30, modified from P. Frost, E. Medina, J.-C. Menaut, O. Solbrig, M. Swift and B. Walker, 1986, 'Responses of savannas to stress and disturbance: a proposal for a collaborative programme of research', *Biology International* (Special Issue 10), Paris (France), International Union of Biological Sciences

[5] Behnke and Scoones, op. cit.

[6] Behnke, Scoones and Kerven, op. cit., adapted from G. Caughley, 1979, 'What is this thing called carrying capacity?' in M. S. Boyce and L. D. Hayden-Wing (eds.), *North American Elk: Ecology, Behaviour and Management*, Laramie,

Wyo., University of Wyoming Press, pp. 2–8, and R. H. V. Bell, 1985, 'Carrying capacity and offtake quotas', in R. H. V. Bell and E. McShane Caluzi (eds.), *Conservation and Wildlife Management in Africa*, Washington DC, US Peace Corps

[7] P. N. de Leeuw and J. C. Tothill, 1993, 'The concept of rangeland carrying capacity in sub-Saharan Africa – myth or reality', in Behnke, Scoones and Kerven, op. cit., pp. 77–88; A. Antenneh, 1984, 'Trends in sub-Saharan Africa's livestock industries', *ILCA Bulletin* (International Livestock Centre for Africa, Addis Ababa (Ethiopia)), 18, 7–15

[8] I. Scoones, 1996, 'New challenges for range management in the 21st century', *Outlook on Agriculture*, 25, 253–6

[9] I. Scoones, 1990, 'Livestock populations and the household economy: a case study from southern Zimbabwe', PhD thesis, University of London; I. Scoones *et al.*, 1996, *Hazards and Opportunities: Farming Livelihoods in Dryland Africa: Lessons from Zimbabwe*, London/New Jersey, Zed Books

[10] J. E. Ellis, M. B. Coughenour and D. M. Swift, 1993, 'Climate variability, ecosystem stability, and the implications for range and livestock development', in Behnke, Scoones and Kerven, op. cit., pp. 31–41

[11] I. Scoones (ed.), 1994, *Living with Uncertainty: New Directions in Pastoral Development in Africa*, London, Intermediate Technology

[12] Behnke and Scoones, op. cit.

[13] I. Scoones, 1993, 'Why are there so many animals? Cattle population dynamics in the communal areas of Zimbabwe', in Behnke, Scoones and Kerven, op. cit., pp. 62–76

[14] ibid.; C. Toulmin, 1994, 'Tracking through drought: options for destocking and restocking', in Scoones, 1996, op. cit., pp. 95–115

[15] N. Alexandratos (ed.), 1995, *World Agriculture: Towards 2010, An FAO Study*, Chichester, Wiley & Sons; FAO, 1993, *Forest Resources Assessment, 1990, Tropical Countries*, Rome (Italy), Food and Agriculture Organization (Forestry Paper 112)

[16] Alexandratos, op. cit.

[17] ibid.

[18] J. Fairhead and M. Leach, 1996a, *Misreading the African Landscape: Society and Ecology in a Forest–Savanna Mosaic*, Cambridge, Cambridge University Press; 1966b, 'Rethinking the forest–savanna mosaic: colonial science and its relics in West Africa', in M. Leach and R. Mearns (eds.), *The Lie of the Land: Challenging Received Wisdom on the African Environment*, Oxford, James Currey/Portsmouth, NH, Heinemann, pp. 105–21

[19] Alexandratos, op. cit.

[20] FAO, 1983, *Fuelwood Supplies in the Developing Countries*, Rome (Italy), Food and Agriculture Organization (FAO Forestry Paper 42)

[21] R. Chambers, N. C. Saxena and T. Shah, 1989, *To the Hands of the Poor: Water and Trees*, New Delhi (India), Oxford and IBH Publishing Co.

[22] Chambers *et al.*, op. cit.; T. Gupta and A. Guleria, 1982, *Some Economic and Management Aspects of a Non-Wood Forest Product in India: Tendu Leaves*, New Delhi (India), Oxford and IBH Publishing Co.

[23] Alexandratos, op. cit.

[24] Ford Foundation, op. cit.; S. B. Roy (ed.), 1995, *Experiences from Participatory Forest Management*, New Delhi (India), Inter-India Publications; M. Hobley and K. Shah, 1996, 'What makes a local organisation robust? Evidence from India and Nepal', *Natural Resource Perspectives*, (Overseas Development Institute, London), No. 11

[25] Alexandratos, op. cit.

[26] Chambers *et al.*, op. cit.

[27] Alexandratos, op. cit.; M. Williams, 1996, *The Transition in the Contribution of Living Aquatic Resources to Food Security*, Washington DC, International Food Policy Research Institute (Food, Agriculture, and Environment Discussion Paper 13)

[28] M. Boonyubol and S. Pramokchutima, 1982, *Trawl Fisheries in the Gulf of Thailand*, Manila (Philippines), International Center for Living Aquatic Resources, (ICLARM Translation 4)

[29] Worldwatch Institute, 1996, *Worldwatch Database 1996*, Washington DC, Worldwatch Institute

[30] World Resources Institute, 1994, *World Resources 1994–1995. A Guide to the Global Environment*, Oxford, Oxford University Press

[31] H. Glantz and J. D. Thompson, 1981, *Resource Management and Environmental Uncertainty: Lessons from Coastal Upwelling Fisheries*, New York, NY, John Wiley; D. Pauly, P. Muck and J. Mendo *et al.* (eds.), 1989, *The Peruvian Upwelling Ecosystem: Dynamics and Interactions*, Callao (Peru), Instituto del Mar del Peru

[32] R. Hilborn and C. J. Walters, 1992, *Quantitative Fisheries Stock Assessment*, New York, NY, Chapman & Hall

[33] D. Ludwig, R. Hilborn and C. Walters, 1993, 'Uncertainty, resource exploitation, and conservation: lessons from history', *Science*, 260: 17, 36

[34] Alexandratos, op. cit.

[35] M. P. Bimbao, A. V. Cruz and I. R. Smith, 1992, 'An economic assessment of rice–fish culture in the Philippines', in W. Hiemstra, C. Reijntjes and E. Van Der Werf (eds.), *Let Farmers Judge*, London, Intermediate Technology, pp. 187–94

[36] K. Kamp, R. Gregory and G. Chowhan, 1993, 'Fish cutting pesticide use', *ILEIA Newsletter*, 2/93, 22–3

[37] B. Maddox, 1994, 'Fleets fight in over-fished waters', *Financial Times*, 30 August 1994 (based on Indrani Lutchman, Fisheries Officer, World Wide Fund for Nature, UK)

[38] B. Holmes, 1996, 'Blue revolutionaries', *New Scientist*, 7 December, pp. 32–5

[39] Williams, op. cit.; FAO, 1992, *Review of the State of World Fishery Resources*, Part 2, *Inland Fisheries and Aquaculture*, Rome (Italy), Food and Agriculture Organization (Fisheries Circular 710 (Rev. 8))

15   Achieving Food Security

Starvation is the characteristic of some people not *having* enough food to eat. It is not the characteristic of there not *being* enough food to eat.

— Amartya Sen, *Poverty and Famines* [1]

The challenge we face, as I have argued in this book, is not simply a matter of meeting the global market demand for food. That is relatively easy. Indeed, it is a goal we have already achieved. Cutbacks in grain production in the developed countries have occurred largely because of a lack of market demand. Yet, as we know, there are over 750 million people in the world who are chronically undernourished. The difficult task over the next twenty-five years is to ensure that they and the millions of others who, on current projections, will be outside the market are well fed. Part of the solution lies in producing more and cheaper food, particularly to feed the urban poor. But if all the poor are to be fed, food production has to increase on the lower-potential lands where a high proportion of them live. And, most important, for the rural landless and for poor families living on insufficient land, food production has to be part of a wider programme of agricultural and natural-resource development that will generate enough employment and income for them to buy the food they need. In effect, food production becomes a means for increasing the market demand for food.

Central to the task is the concept of food security – a concept which is apparently straightforward and amenable to common-sense definition. But, somewhat surprisingly, it is the subject of much debate. A recent review by IDS staff has identified some 200 different definitions [2]. The reason is partly because it is one of those concepts that have entered the political arena, and been repeatedly subject to resolutions at United Nations and other world forums [3]. At the World Food Conference in 1974, food security was defined as 'availability at all times of adequate food supplies of basic food-stuffs . . . to sustain a steady expansion of food consumption . . . and to offset fluctuations in production and prices' [4]. This definition reflected the conference's preoccupation with food shortages in the 1970s. The good harvests of the 1960s, largely a consequence of the Green Revolution, had led to falling prices and to the United States taking land out of production. Both the USA and Canada ran down their grain stores and then in 1972 there were major harvest failures in the Soviet Union, in China, India, Australia and the Sahel. By 1974 grain prices had doubled and there was only enough grain in store to feed the world's population for three and a half weeks [5]. Among the consequences of the conference were the setting up of the World Food Council, the FAO Committee on Food Security and an International Monetary Fund facility enabling countries to meet unexpected needs for food imports.

Despite the seriousness of the situation, this emphasis on food supply as the first priority soon came under challenge, first from nutritionists and nutrition planners, and then from the work of Amartya Sen, who began his seminal essay, *Poverty and Famines*, with the words quoted at the start of this chapter [6]. Sen's studies of the Great Bengal Famine of 1943 had led him to recognize the importance to food security of access to food, as opposed to its supply. Bengal's rice crop had been hit by a cyclone, had suffered from flooding and a disease outbreak, and the Japanese occupation had cut off supplies from Burma. At the time, and subsequently, the famine was attributed to the shortage of food these events created. But Sen's analysis suggested this was only part of the story. The overall shortage of food grains was not that much lower than in previous years, for example 1941, when there had not been a famine. In Sen's view the evidence suggests it was not a 'remarkable' shortfall. More important, in the rural areas, where the famine hit hardest, was the failure of labourers' wages to keep pace with the inflation induced by the war economy. Well before the famine the price of rice had doubled, but wages had risen little (Figure 15.1). A major cause of the famine was the inability of the rural poor to purchase the rice they needed. They migrated in large numbers to Calcutta, where relief measures proved inadequate, despite the availability of sufficient food. In all, some 3 million people died.

His study of this and other subsequent famines in Ethiopia, the Sahel and Bangladesh led him to develop the concept of 'entitlement'. Put simply, people are entitled to food because they have produced it themselves

286

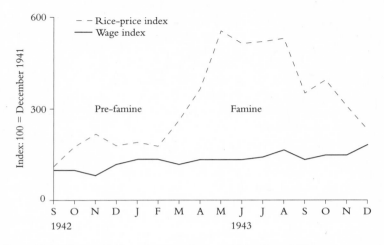

Figure 15.1 Wages and rice prices in a district of West Bengal [7]

– as farmers or sharecroppers – or they have earned the money to buy it or they can receive food as part of an exchange with their neighbours or kin, or under a government system of benefit. This emphasis on access to food rather than food production produced a sea change in thinking about food security [8]. FAO, in a 1983 reappraisal, stressed 'physical and economic access to . . . basic food' [9] and in 1986 the World Bank adopted the following definition: 'Food security is access by all people at all times to enough food for an active, healthy life' [10].

However, as Simon Maxwell of IDS argues, this multiplicity of definition reflects not only these political policy changes, but also the great diversity of people's own, subjective perceptions of their food-security status [11]. Research over the last decade in famine situations has revealed the complex ways in which people respond to adversity. Often their actions seem, to the outsider, counter-intuitive. Alex de Waal, of Oxford's Queen Elizabeth House, working during the 1984/5 famine in Darfur in the western Sudan, observed that people did not place reducing immediate hunger top of their list of priorities [12]. They preferred, instead, to keep their seed for planting, not to sell their animals and to expend time and resources on cultivating their fields. In effect, they gave first priority to preserving their assets for the future. What this and other observations indicate is that food security for the household or the individual is but one part, albeit a crucial part, of a livelihood strategy and needs to be analysed and understood in this context.

In this, penultimate, chapter I discuss food security at three levels – the global, the national and the household (including the individual) – at each level examining whether food is available and, if so, whether it is affordable or is otherwise accessible.

Over the past thirty years there has been a massive expansion of world trade in food and other agricultural products. In monetary terms, cereals are now second only to petroleum in international trade [13]. Part of the increase in trade to date has been to satisfy the former Soviet Union's demand for cereals, but a major factor has been increasing incomes in East Asia and parts of West Asia and North Africa, resulting in growing demands for livestock products and hence for grain as feed. Over 40 per cent of the volume of global cereal trade is in maize and other coarse grains. Yet, for many developing countries imported food has become less affordable. Alongside the increase in cereal imports, the agricultural exports of the developing countries have steadily declined as a proportion of world trade. And for those dependent on the export of agricultural commodities, such as coffee, tea and sugar, the decline in their prices has made it more difficult for them to finance their food imports. The developing countries were net exporters of agricultural products in the 1960s but by 1992 they had become net importers (Figure 15.2) [14].

Compounding this general trend are the considerable variations in grain prices which relate, in a rather complex manner, to fluctuations in grain production and the level of grain stocks [15]. In 1970 stocks were high (equivalent to about 80 days of global consumption) but fell in 1973, following the steep rise in grain prices (Figure 15.3). Stocks were again high in the early 1980s and the USA withdrew over 30 million hectares of crop land from grain production in 1983 (see Chapter 7), but this was followed by serious drought in the USA and elsewhere and stocks once again fell. There was another peak in 1987 and then a decline. In 1995 stocks were 295 million tons, equivalent to only 61 days of consumption, and the 1996 estimate is 229 million tons, equivalent to a mere 48 days, at the lower end of the range FAO considers the minimum to safeguard world food security.

These severe fluctuations are partly the result of climatic conditions,

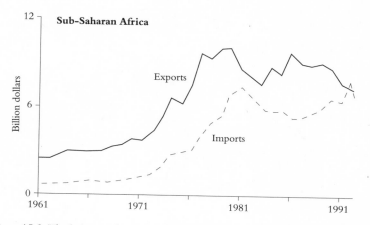

Figure 15.2 The balance of agricultural imports and exports in Sub-Saharan Africa (current prices)

but also of market responses and changes in governmental policy. An example is the behaviour of the global wheat market over the last ten years (Figure 15.4). Production grew steadily in the 1980s, reaching a peak of nearly 600 million tons in 1990. It then slowly declined, followed by a sudden drop owing to a fall in production in 1994 in eastern Europe and the former Soviet Union, and in the following year to adverse weather in the USA and in a number of other major exporters, Spain, Argentina and Australia. This was partly compensated for by record harvests in East Europe in 1995, but the Russian harvest was the worst for thirty years and, overall, wheat production fell below demand in 1995. Wheat prices rose to a new high ($200/ton in current prices) and by 1996 wheat stocks had fallen to 88 million tons. According to FAO, the developing countries had to pay $3 billion for their grain imports, an increase of 25 per cent over the previous year. The response of the European Union was to cease its policy of export subsidies, to impose a tax on wheat exports and to reduce the amount of land in set-aside. The USA also eliminated export subsidies and, for 1996, some of the Conservation Reserve land came back into production. Wheat planting grew by 4 per cent in the European Union and worldwide the amount planted was the greatest since

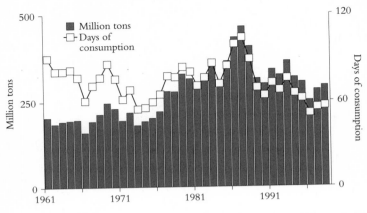

Figure 15.3 World grain stocks and their equivalent in terms of days of global consumption (figures are for year when new harvest begins; 1997 is an estimate) [16]

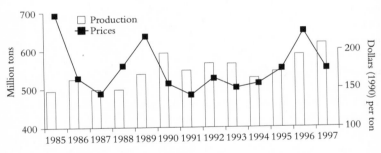

Figure 15.4 The global wheat market in the 1990s (US export prices) [17]

1986. Yields in the EU are estimated to be nearly 8·25 tons/ha and estimated global production will be over 550 million tons. Prices fell rapidly towards the end of 1996 and the EU was once again subsidizing exports.

How can the adverse terms of trade be reversed and the fluctuations be avoided or, at least, minimized? In 1993, after eight years of negotiations,

115 countries (the majority developing countries) concluded the Uruguay Round of the Multilateral Trade Negotiations [18]. Its aim is to create a more liberalized trade in industrial and agricultural products. Under the Agreement on Agriculture there are to be programmed reductions in domestic support, in export subsidies and in tariffs and other limitations on market access. In theory, this should benefit farmers in the developed and developing countries alike. By reducing the protection afforded developed-country farmers, there will be less food exported at prices which undercut developing-country farmers. And because import barriers are to be reduced, there should be growing demand for agricultural products (staples and non-staples), leading to increased agricultural production in both developed and developing countries.

However, the changes will occur only slowly and the benefits to the developing countries in the short term are not obvious. The prices of their own agricultural commodities for export will rise little, if at all, and will continue to be highly volatile. International commodity agreements have not worked and the developing countries will have to try to mitigate the variability and the long-term decline in the prices of their main export goods through trading techniques such as long-term contracts at fixed prices and by drawing on the compensatory financing facilities of the IMF [19]. Most commentators also believe the Uruguay Round will increase the prices of developed-country products by 5–10 per cent. This will mean higher costs for food imports, a rise of over $3 billion (or 15 per cent) for the developing countries as a whole, according to the FAO [20].

The Uruguay Round's reductions in domestic support and eventual price stabilization could also spread the effects of food-price fluctuations across a larger number of countries. But this will not occur if stocks remain low. Government food stocks are predicted to decline, and there are doubts whether they will be replaced by an increase in stocks held by the private sector [21]. The removal of sudden policy changes by developed-country governments should be beneficial, but production will continue to fluctuate, primarily as a result of vagaries in the climate and speculation by developed-country growers and market traders.

For countries in trouble, there is always food aid. Although the USA and the UK were providers of aid to a number of Asian, African and Latin American countries in the last century and the early part of this century, the modern form of food aid owes much to the Marshall Plan at the end of the Second World War, through which some $13·5 billion of aid (25 per cent in the form of food, feed and fertilizer) was provided to war-ravaged Europe by the USA and Canada [22]. The food was sold by the recipient countries for local currency, so freeing their hard currency for other imports. Although in essence humanitarian, the Plan had mutual benefits for donors and recipients: the former had grain surpluses they needed to dispose of and farmers whose livelihoods required protection, while the recipient countries were able to feed their populations (in the immediate post-war years 125 million Europeans were averaging only 2,000 calories per day) and rebuild their industries.

The same principles have continued under the aid programmes to the developing countries, particularly the USA's PL480 programme and those of the European Union. However, unlike the situation at the time of the Marshall Plan, when recipient countries were rehabilitating their industries, the need in the developing countries is for infrastructural and other underpinnings of longer-term development. To some extent, the linkage of aid to development was addressed by the creation of the World Food Programme in 1962, administered by FAO and designed to be freer from some of the political constraints faced by individual donor countries. But in practice, food aid comes in many forms and with a variety of objectives. Most is cereals, but edible oils, dairy products and fish are also important. Nearly a quarter is emergency aid, meeting the immediate requirements of populations affected by natural disasters, civil conflict and war. Other aid goes towards support of individual projects or programmes, although the distinction between different forms of aid is increasingly blurred, in part because the recurrence of disasters in certain countries makes programmed food aid more appropriate. The amounts of aid were at their highest in the 1960s, when approximately half of all developing-country cereal imports were in the form of aid. There were subsequent, smaller peaks in the early 1970s and mid-1980s (Figure 15.5). Most striking today is the rising proportion of aid going to Sub-Saharan Africa – but overall food aid has levelled off.

Its contribution to food security is also being questioned. Because the volume of aid continues to be linked to the level of surplus stocks, it is highest when food is most available and lowest when most needed. Thus at the global level it enhances rather than dampens the grain cycles. And as Edward Clay and Olav Stokke point out, the amounts are too small to be significant for a country such as India, and are likely to be limited even

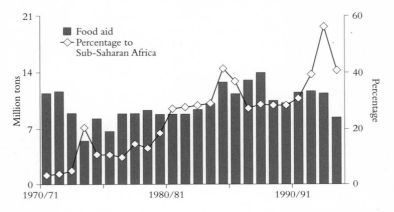

Figure 15.5 Food aid received in the developing countries and proportion going to Sub-Saharan Africa [23]

for the needs of Bangladesh [24]. In the 1988/9 flood, Bangladesh received 1·6 million tons of cereals in the form of aid and paid for a further 2·2 million tons. If food aid continues to decline, as stocks are run down following the Uruguay Round, say to the 9 million tons level of the early 1980s, even 1·6 million tons might be difficult to provide and a requirement of two to three times this amount would be impossible.

For the future, food aid will be limited to countries such as those in Sub-Saharan Africa where the import requirements, although significant for the country, are small with respect to overall world trade. Even there, as the widespread drought of the 1980s showed, much has still to be learnt if relief is to be speedy and result in a sustainable level of security. There are real dangers of aid destroying local incentives, for example driving down local prices and shifting consumption away from indigenous staples, as in the Sudan, where preferences moved from locally grown sorghum to bread made from subsidized wheat. Returns to local farmers, in the face of competition from imports, are likely to be reduced, impeding long-term agricultural development [25].

To some extent, national attempts to improve food security mirror those at the global level [26]. Concerted efforts are being made to increase production and, for most countries, this means aiming at market self-sufficiency, at least in grain production, and the establishment of stocks for emergencies and systems of food distribution at times of need. In the poorest countries these goals have been set over the past decade within overall programmes of economic reform and structural adjustment [27]. Loans from the World Bank and the International Monetary Fund have been conditional on the adoption of policies aimed at creating full market economies. Public spending has been reduced and exchange rates devalued to encourage higher investment and exports.

The effectiveness of these policies is a matter for debate. According to studies by Uma Lele and Hans Binswanger of the World Bank, agricultural declines have been halted in several countries and there have been significant increases in agricultural growth in others [28]. Ghana undertook a range of reforms beginning in 1983, which included a 90 per cent exchange-rate devaluation and a progressive rise in the price of the main export crop, cocoa [29]. As a result cocoa exports, which had fallen from a high of 450,000 tons in the 1970s to only 150,000 tons in 1983, have partially recovered and now stand at 250,000 tons. But because cocoa prices have still remained relatively low, the benefits have been limited. For Ghana and other similar countries there is an urgent need to diversify their agricultural exports and become more flexible, enabling them to take better advantage of market opportunities.

Most commentators agree that, whatever the long-term benefits of structural adjustment, the short-term social costs can be considerable. Unemployment may increase, particularly as a result of reduced public expenditure, and this will result in greater poverty and less food security among some sections of the community. Mali, as part of its programme of economic reforms begun in 1982, shed several thousand civil-service jobs [30]. A family member in the civil service had been a traditional insurance for rural people, guaranteeing at least some cash income in bad years, and the loss of the jobs created widespread food insecurity.

'Getting the prices right' is rarely, by itself, sufficient to increase production and incomes. Where agriculture is dominated by low production technologies and served by poor infrastructure, price incentives have not been enough to make a difference. The comparison between the poorest countries of Asia and Sub-Saharan Africa is instructive. Economic reforms have been more successful in Asia partly because they were based on more developed financial systems, stronger private markets and indigenous rural institutions, and better infrastructure. Asia has more developed road and rail networks linking farmers with their markets (Figure 15.6). Public expenditure on agricultural extension and research is also required if farmers are to take advantage of the new opportunities [31].

And, as the Mali experience of food-security programmes clearly shows, the market also requires supportive government planning that needs to be sensitive to regional and local variations [32]. Mali, like most developing countries, contains areas of high and lower potential that have different ecological and socio-economic conditions and these require differing approaches to the achievement of food security. As Simon Maxwell of IDS puts it: 'We need a new food security policy . . . "Walking on two legs, but with one leg longer than the other" ' [33]. The 'longer leg' is directed at the areas of lower potential where the problems are generally more complex, the need is greater and there is, currently, a lack of public investment. There is less room for macro-solutions − innovations and investments that are replicable on a wide scale. Instead, there is a requirement for:

− Appropriate technologies, aimed at local conditions, and aimed at the needs of households and individuals;

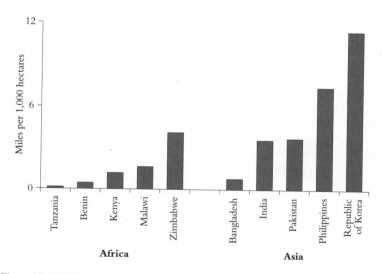

Figure 15.6 Miles of road and rail per 1,000 hectares of cultivable area [34]

– The development of local institutions; and
– Supportive actions by local government agencies and non-government organizations.

Working at this level can, in turn, provide insights and information that will hopefully direct international and national policies along better lines.

According to conventional wisdom, basic human needs can be arranged in a strict hierarchy; people strive to satisfy one need before moving on to the next. In the hierarchy, food is assumed to be the need which has to be met first. But we now know that, except in a trivial sense, this is far too simplistic a model. All people, whether they are rich financiers or poor farmers, take both short- and long-term decisions – 'they juggle', in the words of Susanna Davies of IDS, 'between immediate consumption and future capacity to produce' [35].

We have gained a much better understanding, in recent years, of how poor people behave in the face of adversity. This has come from studies of famine situations and from the experience of aid workers attempting to mitigate famines while ensuring that immediate relief is tied to longer-term development [36]. In Susanna Davies's study of the effects of drought on rural households in Mali in the 1970s she describes a complex cycle of responses (Figure 15.7). In the first two years of the drought the food stocks are run down, then surplus cattle are sold. Various coping strategies are tried in the third year: they collect wild foods or temporarily migrate elsewhere. In the next year they may have to borrow food from their kin, but in subsequent years, as the rainfall improves, they pay back the loans and reinvest in cattle.

Such behaviour can only be understood in terms of poor people's livelihood strategies. As I indicated in Chapter 9, livelihoods are complex phenomena consisting of people, their activities and their assets, both tangible and intangible (Figure 15.8).

Tangible assets include:

– Stores: food stocks, items of value such as gold, jewellery and woven textiles, and cash savings in banks or thrift and credit schemes; and
– Resources: land, water, trees, livestock, tools and other farm equipment.

Intangible assets include:

– Claims: the demands and appeals which can be made on neighbours or

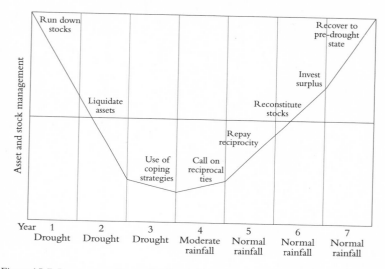

Figure 15.7 Responses of households to drought in Mali [37]

communities or agencies for support and for access to stores and resources; and

− Access: the opportunity to use a resource or store or a service, such as transport, education, health and markets, or to obtain information and technology.

Out of these assets people construct and contrive a living using their physical labour, their skills, knowledge and creativity. In general, the richer the assets and the more labour and knowledge they have, the better is the living. But, inevitably, people have to make trade-offs, of the kinds I described in Chapter 9, and, in particular, between immediate consumption and longer-term sustainability. At the heart of these trade-offs is security, not just of food, but of income, health, status, and of less measurable attainments, such as freedom of belief and expression, and of 'peace of mind'. Security is about both reality and perception. The head of a household needs assurance that there will be enough food, available and accessible, for his or her household now and into the foreseeable future.

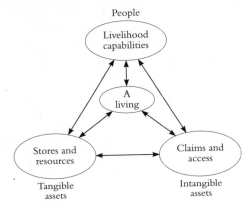

Figure 15.8 Components and flows in a livelihood [38]

Households will inherit and accumulate assets and will draw on these at times of adversity. But if, as has happened more recently in Mali, droughts are of longer duration and assets have been drawn down to the point at which households find it difficult to recover, then outside intervention is crucial to save not only lives but also livelihoods. This may take the form of food donations, provision of short-term credit or of seed for sowing, or the creation of opportunities for off-farm employment [39].

Once basic survival is assured, and conditions are secure, farmers will replace and improve their assets. They may enhance resources by using their labour to construct terraces which will improve the stock of soil, or by investing in a cart to take produce to market. Claims may be established by investing in a marriage or by giving presents. Access to information may be obtained by purchasing a radio. Capabilities can be increased by education or apprenticeship.

Some of these investments will be aimed at increasing productivity, some at reducing instability, some at increasing sustainability – in the sense of reducing the liability to shock and stress – and some may increase equitability – the sharing of benefits within the household. We can see some of these investments at work in the intensive farm at Kakamega, in Kenya, which I described in Chapter 2. There the head of the household, a woman, has created a very diverse plot of land on which the annual

298

crops are interspersed with trees. The trees on her farm, and neighbouring farms, are essentially a kind of subsistence crop, being sold to obtain basic food during the lean months of the year, or are savings for the inevitable bad year or the expense of a wedding or funeral, or are grown for cash, in particular for the regular payment of school fees. In the latter case the sustainable investment is in the long-term future of the children.

Part of the clue to this remarkable pattern of investment is the security of ownership of their land. Security of access to land is usually a necessary, although not a sufficient, condition for investment in land improvement. Sometimes it may take the form of security of ownership or tenure, or it may be more complicated. In the Joint Forest Management programme in West Bengal, described in the previous chapter, the willingness of villagers to manage the forest on a sustainable basis depends on a lease negotiation with the government. And the long-term management of their livestock by the pastoralists in Zimbabwe depends on a complex pattern of traditional rights which extend over a range of habitats, some of which may be only rarely exercised. Security of access must also extend to the produce. Ownership of land may not be sufficient to encourage tree-planting or terracing if other rights are not also secured or other conditions are not satisfied. As I mentioned in the previous chapter, Indian government prohibitions on the felling of trees, even on smallholders' land, actively discourage tree-planting.

The Kakamega farm also illustrates the general point that a diversity of productive activity helps achieve food security. Diversity can come from growing a range of crops and tending a variety of livestock, but most farm households in the developing countries also benefit from off-farm income. Farmers in upland Java depend heavily on carpentry, trade and the selling of charcoal, palm-sugar and wood, as well as on wage labour [40]. For farms of less than half a hectare this can make up 95 per cent of the cash income. In these circumstances, potential agricultural innovations have to take into account the improved returns to agriculture relative to those from off-farm labour.

In addition to locally earned income, very large sums of money are often sent back to rural households by individuals who have temporarily or permanently migrated to the major urban centres or even overseas [41]. Rural livelihoods in Mexico have significantly benefited through such remittances from the United States and, similarly, in the southern African nations through remittances from South Africa, and in many Asian countries from Saudi Arabia and the Gulf States. Even quite remote villages in the mountains of Pakistan or small islands in the Philippines are tied in this way to the economies of the oil-rich Arab states. The total amounts involved in worldwide remittances are not known; they probably exceed official development assistance. More important, very little is yet known about how such remittance moneys are used. To what extent are they spend on food, on consumer goods, on education, or on investment in land and long-term agricultural production?

Despite often significant sources of off-farm income, lack of credit is a major constraint to investment for most farmers and hampers their ability to survive adverse conditions. The government credit schemes set up in the early years of the Green Revolution have proven effective in the high-potential lands where high and fairly reliable returns to investments can be achieved. More problematic has been the provision of credit in the lower-potential areas, and particularly for the poorest households, where the returns are low and the risks high. Typically, such loans are very small, yet require careful attention and hence are relatively costly to service. They are not attractive to commercial banks and are difficult for more bureaucratic government agencies to handle. The traditional alternative is the moneylender, often charging very high rates of interest that create long-term and increasing indebtedness.

However, there is now sufficient experience, going back over two decades, to demonstrate the effectiveness of local, self-managed credit groups. The key to the formation of such groups is often a coming together of households on a small development activity that satisfies a collective need [42]. In the valleys of northern Pakistan, the Aga Khan Rural Support Programme (AKRSP) began by helping the local communities to organize around the construction of new irrigation channels or marketing roads (see Chapter 10). These village organizations then became the focus of saving schemes. Once the savings had grown, the organizations were able to obtain complementary loans from commercial banks, the AKRSP acting as a partial guarantor. By 1994, over 50,000 households were members of such organizations [43].

In southern India a similar, although somewhat more informal, approach has been developed by an NGO, MYRADA [44]. There the savings and credit groups are initially organized around a small project such as desilting a tank, or creating a drainage system. In one case the group started from digging a trap to catch a marauding elephant. The collective physical activity and the experience of cooperative planning and management in the project lay the basis for trust and self-confidence. Although there are guidelines, each scheme tends to evolve along rather different lines, reflecting its origin and local circumstances. Members determine the rules, set interest rates, decide on the types of loans and vet and approve the loan applications. Most of the loans are very small, less than 100 rupees ($3), and are used to pay for a marriage or funeral, to buy food ahead of harvest or to purchase one or two sheep. By 1992 some 108 million rupees ($3·6 million) had been dispersed to some 50,000 members of these groups.

Perhaps the best known of all credit schemes is the Grameen Bank, established in Bangladesh in the 1970s and now encompassing over 1·5 million members [45]. At the beginning of this book I described the story of Koituri, a Bangladeshi woman who had left her husband to escape his constant beatings. She had moved to her father's compound and worked at various tasks for better-off neighbours, receiving two meals a day and some rice. Her life began to improve when the Grameen Bank opened a nearby branch. Some women formed groups in her village. At first she was nervous, hearing stories of arrests and imprisonment if loans were not repaid. But eventually she joined a group of nine women and received training from the bank. She took out a small loan to buy a goat and a kid. She tethered the goat where she worked and cut grass to feed it. But it did not prosper and she sold it for a loss. However, with the money she leased a tiny plot and planted lima beans on bamboo scaffolding. These she successfully sold; she took a new loan, rented some more land, bought a plough, paid for irrigation and labour and began to grow an IRRI rice variety. Eventually she paid off all her loans, acquired a new loan, with which she bought a calf, and when Mohammed Yunus was writing about her she was contemplating a loan to replace her hut with a small house. In Yunus's words:

The bank loan has not only changed her financial status, it has also brought about a change of her outlook. She used to feel frightened all the time before . . . Now she does not feel insecure any more. She has her friends from the group to look after her. Previously, she worked as a maidservant to others and was treated like one. Now she has money in her hand. People come to her in their trouble and she can help them with small loans. [46]

In the global scale of affairs it is a small and insignificant achievement, but for her it has meant the difference between life and death, and brought dignity and self-confidence. If the experience was multiplied for the poor and hungry throughout the developing world it would add up to a revolution.

Notes

[1] A. Sen, 1981, *Poverty and Famines: An Essay on Entitlement and Deprivation*, Oxford, Clarendon Press

[2] M. Smith, J. Pointing and S. Maxwell, 1992, *Household Food Security, Concepts and Definitions: An Annotated Bibliography*, Brighton (UK), Institute of Development Studies, University of Sussex (Development Bibliography No. 8)

[3] S. Maxwell, 1996a, ' "Walking on two legs, but with one leg longer than the other": a strategy for the world food summit', *Forum Valutazione*, 9, 89–103; 1996b, 'Food security: a post-modern perspective', *Food Policy*, 21, 155–70

[4] United Nations, 1975, *Report of the World Food Conference*, Rome (Italy)

[5] C. Tudge, 1977, *The Famine Business*, London, Faber & Faber

[6] A. Sen, 1976, 'Famines as failures of exchange entitlements', *Economic and Political Weekly*, 11, 1273–80 (Special Number); 1977, 'Starvation and exchange entitlements: a general approach and its application to the Great Bengal Famine', *Cambridge Journal of Economics*, 1, 33–60

[7] Sen, 1981, op. cit.

[8] Maxwell, 1996b, op. cit.

[9] FAO, 1983, *World Food Security: A Reappraisal of the Concepts and Approaches*, Rome (Italy), Food and Agriculture Organization (Director-General's Report)

[10] World Bank, 1986, *Poverty and Hunger: Issues and Options for Food Security in Developing Countries*, Washington DC, World Bank

[11] Maxwell, 1996b, op. cit.

[12] A. de Waal, 1989, *Famine That Kills*, Oxford, Clarendon Press; 1991, 'Emergency food security in Western Sudan: what is it for?' in S. Maxwell (ed.), *To Cure All Hunger: Food Policy and Food Security in Sudan*, London, Intermediate Technology

[13] D. Grigg, 1993, *The World Food Problem* (2nd edn), Oxford, Blackwell

[14] N. Alexandratos (ed.), 1995, *World Agriculture: Towards 2010. An FAO Study*, Chichester (UK), Wiley & Sons

[15] I. Dyson, 1996, *Population and Food: Global Trends and Future Prospects*, London, Routledge

[16] L. R. Brown, M. Renner and C. Flavin, 1998, *Vital Signs 1998: The Environmental Trends That are Shaping Our Future*, New York, NY, W. W. Norton, using data from USDA, 1997, *Grain: World Markets and Trade*, January, Washington DC, United States Department of Agriculture, and USDA, 1998, *Production, Supply and Demand View* (electronic database), February, Washington DC, United States Department of Agriculture

[17] World Bank, 1998, Washington DC, World Bank (Development Economics Prospects Group); USDA, 1998, *Wheat Situation and Outlook Yearbook*, Washington DC, United States Department of Agriculture

[18] Alexandratos, op. cit.; FAO, 1995, *The State of Food and Agriculture, 1995. Agricultural Trade: Entering a New Era*, Rome (Italy), Food and Agriculture Organization

[19] Alexandratos, op. cit.

[20] FAO, 1995, op. cit.

[21] FAO, 1995, op. cit.; N. Islam, 1996, 'Implementing the Uruguay Round: increased food stability by 2020', *2020 Brief*, 34, Washington DC, International Food Policy Research Institute

[22] H. Singer, J. Wood and T. Jennings, 1987, *Food Aid: The Challenge and the Opportunity*, Oxford, Clarendon Press

[23] FAO, 1983, 1994, *Food Aid in Figures*, Rome (Italy), Food and Agriculture Organization

[24] E. Clay and O. Stokke, 1991, 'Assessing the performance and economic impact of food aid: the state of the art', in E. Clay and O. Stokke (eds.), *Food Aid Reconsidered: Assessing the Impact on Third World Countries*, pp. 1–36, London, Frank Cass (EADI Series 11)

[25] S. Maxwell, 1991, 'The disincentive effect of food aid: a pragmatic approach', in Clay and Stokke, op. cit., pp. 66–90

[26] N. Islam and S. Thomas, 1996, *Foodgrain Price Stabilization in Developing Countries: Issues and Experiences in Asia*, Washington DC, International Food Policy Research Institute

[27] Alexandratos, op. cit.

[28] U. Lele, 1992, 'Structural adjustment and agriculture: a comparative perspective of performance in Africa, Asia and Latin America', *29th Seminar of the European Association of Agricultural Economists,* Hohenheim (Germany); H. Binswanger, 1989, 'The policy response of agriculture', *Proceedings of the World Bank Conference on Development Economics*, Washington DC, World Bank

[29] World Bank, 1986, *Financing Adjustment with Growth in Sub-Saharan Africa, 1986–1990*, Washington DC, World Bank; W. Seini, J. Howell and S. Commander, 1987, 'Agricultural policy adjustment in Ghana', *Conference on the Design and Impact of Adjustment Programmes on Agriculture and Agricultural Institutions, September 10–11, 1987*, London, Overseas Development Institute

[30] S. Davies, 1996, *Adaptable Livelihoods: Coping with Food Insecurity in the Mahelian Sahel*, London, Macmillan Press/New York, NY, St Martin's Press

[31] T. W. Schultz, 1990, 'The economics of agricultural research', in C. Eicher and J. Staatz (eds.), *Agricultural Development in the Third World*, Baltimore, Md, Johns Hopkins University Press

[32] Maxwell, 1996a, op. cit.

[33] ibid.

[34] Alexandratos, op. cit. using data from J. Ph. Platteau, 1993, 'Sub-Saharan Africa as a special case: the crucial role of (Infra) structural constraints', *Cahiers de la Faculté des Sciences Économiques et Sociales de Namur* (Série Recherche No. 128, 1993/6), and R. Ahmed and C. Donovan, 1992, *Issues of Infrastructural Investment Development. A Synthesis of the Literature*, Washington DC, International Food Policy Research Institute

[35] Davies, op. cit.

[36] Maxwell, 1996b, op. cit.

[37] Davies, op. cit.

[38] R. Chambers and G. R. Conway, 1992, *Sustainable Rural Livelihoods: Practical Concepts for the 21st Century*, Brighton (UK), Institute of Development Studies, University of Sussex (Discussion Paper No. 296)

[39] Davies, op. cit.

[40] P. Van den Poel and H. Van Dijk, 1987, 'Household economy and tree-growing in upland Central Java', *Agroforestry Systems*, 5, 169–84

[41] G. R. Conway and E. B. Barbier, 1990, *After the Green Revolution: Sustainable Agriculture for Development*, London, Earthscan

[42] J. N. Pretty, 1995, *Regenerating Agriculture: Policies and Practice for Sustainability and Self-reliance*, London, Earthscan

[43] AKRSP, 1994, *Annual Report*, Gilgit (Pakistan), Aga Khan Rural Support Programme

[44] A. Fernández, 1992, *The MYRADA Experience: Alternative Management Systems for Savings and Credit of the Rural Poor*, Bangalore (India), MYRADA; V. Ramaprasad and V. Ramachandran, 1989, *Celebrating Awareness*, Bangalore (India), MYRADA/New Delhi (India), Foster Parents Plan International

[45] M. Hossain, 1988, *Credit Alleviation of Rural Poverty: the Grameen Bank in Bangladesh*, Washington DC, International Food Policy Research Institute (IFPRI Research Report 65); P. S. Jain, 1996, 'Managing credit for the rural poor: lessons from the Grameen Bank', *World Development*, 24, 79–89

[46] M. Yunus (ed.), 1984, *Jorimon of Beltoil Village and Others: In Search of a Future*, Dhaka (Bangladesh), Grameen Bank

16   After the World Food Summit

We pledge our political will and our common and national commitment to achieving food security for all and to an ongoing effort to eradicate hunger in all countries, with an immediate view of reducing the number of undernourished people to half their present level no later than 2015.

            — *The Rome Declaration on World Food Security* [1]

In November of 1996 the FAO convened a World Food Summit at its headquarters in Rome. It was billed as a meeting of heads of state although, in the event, it was mostly attended by aid ministers from the developed countries and by agriculture ministers from the developing countries. Because the declaration and plan of action had been drafted and agreed beforehand, the meeting was largely an opportunity for speech-making. In some respects, it was simply one more international summit in a sequence which included the World Summit for Children in New York in 1990, the United Nations Conference on Environment and Development in Rio de Janeiro in 1992, the International Conference on Population and Development in Cairo in 1994, the World Summit for Social Development in Copenhagen, and the Fourth World Conference on Women in Beijing, both in 1995.

There is considerable cross-referencing to these earlier summits in the World Food Summit declaration, rightly pointing out the interconnectedness of their concerns with food security. Like most of the previous meetings it committed governments to a long list of general objectives but, with the exception of the target quoted at the beginning of this chapter, was short on specific objectives.

By way of implementation, FAO has proposed a global 'Food for All Campaign', based on national committees representative of government, the private sector and civil society, and a special programme aimed at increasing food production in Low-Income Food-Deficit Countries

(LIFDCs) [2]. The latter is the more significant and will be crucial if the summit's aim of reducing the number of chronically undernourished by half is meant seriously. FAO lists over seventy-five countries as Low-Income Food-Deficit (Box 16.1). Together they contain 3·5 billion people, nearly two-thirds of the world's population. Most severely affected are the thirty-five countries (twenty-five in Sub-Saharan Africa) whose average per capita daily calorie intake is less than 2,200 calories. In total they contain over 675 million people, a great majority of whom are chronically under-nourished; a programme that increases their average calorie availability by 200/300 calories per day would have a significant effect on global hunger, but the difficulties of achieving the FAO target within a mere fifteen years should not be underestimated. Many of these countries have been racked by civil war, many have unstable or ineffective governments characterized by rundown infrastructure, poor transportation links, inadequate markets, a lack of appropriate food policies and weak agricultural extension and research. FAO has begun a pilot programme in fifteen of these countries, although the details have not yet been published [3].

Where I believe the summit was particularly successful was in its conceptual thinking. It acknowledged, in the declaration, the multifaceted character of food security. In this, it departed in a radical manner from the pronouncements of earlier meetings, notably the first World Food Conference in 1974, where most of the emphasis was on population control and on food production in the high-potential areas. Many of the statements in the Declaration and Plan of Action echo those in the report of the CGIAR's Vision Panel produced in 1994 and are in agreement with the themes of this book [4]. The Declaration and Plan emphasize the importance for food security of eradicating poverty and inequality and, while not referring to entitlements as such, stress the improvement of 'physical and economic access by all, at all times, to sufficient, nutritionally adequate and safe food and its effective utilization' [5]. Sustainability is a common theme and, most important, there is a clear acknowledgement of the need to develop both high- and low-potential areas and for a participatory approach to development. These views are perhaps best summed up in Objective 3.5 of the Plan of Action:

To formulate and implement integrated rural development strategies, in low and high potential areas, that promote rural employment, skill formation, infrastructure, institutions and services, in support of rural development and household food

306

Box 16.1 **Low-Income Food-Deficit Countries**
All food-deficit countries with per capita incomes below $1,395 per annum,
divided into those with average per capita calorie supplies of less than and
more than 2,200 per day in 1994 (population size, in millions, in parentheses) [6]

LESS THAN 2,200 CALORIES	MORE THAN 2,200 CALORIES

East Asia/Pacific

Cambodia (12·9) | China (1,208·8) | Philippines (66·2)
Mongolia (2·4) | Indonesia (194·6) | Samoa (0·2)
Solomon Islands (0·4) | Kiribati (0·1) | Tuvalu (0·01)
| Laos (4·7) | Vanuatu (0·2)
| Papua New Guinea (4·2) |

South Asia

Afghanistan (18·9) | Bhutan (1·6) | Pakistan (136·6)
Bangladesh (117·8) | India (918·6) | Sri Lanka (18·1)
Nepal (21·4) | Maldives (0·2) |

West Asia/North Africa

Yemen (13·9) | Egypt (61·6) | Morocco (26·5)
| Jordan (5·3) | Syria (14·1)

Latin America/Caribbean

Bolivia (7·2) | Colombia (34·5) | Guatemala (10·3)
Haiti (7·0) | Dominican Republic (7·7) | Honduras (5·5)
Peru (23·3) | Ecuador (11·2) | Nicaragua (4·3)
| El Salvador (5·6) |

Sub-Saharan Africa

Angola (10·7) | Mozambique (15·5) | Benin (5·2) | Gambia (1·1)
Burundi (6·2) | Nigeria (108·5) | Burkina Faso (10·0) | Guinea (6·5)
Cameroon (12·9) | Rwanda (7·8) | Cape Verde (0·4) | Guinea-Bissau (1·1)
Central African Republic (3·2) | São Tomé & Príncipe (0·1) | Congo (2·5) | Lesotho (2·0)
Chad (6·2) | Sierra Leone (4·4) | Côte d'Ivoire (13·8) | Mali (10·5)
Comoros (0·6) | Somalia (9·1) | Djibouti (0·6) | Mauritania (2·2)
Ethiopia (53·6) | Sudan (27·4) | Equatorial Guinea (0·4) | Niger (8·8)
Ghana (16·9) | Tanzania (28·8) | Eritrea (3·4) | Senegal (8·1)
Kenya (27·3) | Uganda (20·6) | | Swaziland (0·8)
Liberia (2·9) | Zaire (42·6) | | Togo (4·0)
Madagascar (14·3) | Zambia (9·2) |
Malawi (10·8) | Zimbabwe (11·0) |

307

security and that reinforce the local productive capacity of farmers, fishers and foresters and others actively involved in the food sector, including members of vulnerable and disadvantaged groups, women and indigenous people, and their representative organizations, and that ensure their effective participation [7].

This is an objective which I strongly endorse yet, as I have argued in this book, I believe it can only be achieved if we embark on a new, well-funded international programme of research, on a scale and with a vision at least equal to that of the Green Revolution.

The Green Revolution was successful not only because of the quality and relevance of the science and technology. Leadership, skilled management and the creation of the right institutional settings were crucial. At the International Rice Research Institute (IRRI), the wheat and maize centre (CYMMIT) and the other international research institutes, well-resourced environments were created in which teams of scientists could work effectively together towards clearly defined goals. Plant breeders, geneticists, agronomists, plant pathologists and entomologists collaborated on the development of the new high-yielding varieties. And the production of the packages accompanying the delivery of the new seeds to the farmers depended on close interaction between the research scientists and the extension specialists.

Success was not easily achieved. The goals we now face are, in many respects, more complex. It is no longer sufficient to focus solely on producing new high-yielding varieties, important though that is, or the design of packages of inputs. The relatively uniform, highly favoured, high-potential Green Revolution lands are now not the only target for innovative research and implementation. There is a diversity, in both agro-ecological and socio-economic terms, of targets and problems to be tackled. To the goal of increased productivity we have added equally important objectives of sustainability, stability and equity, and explicitly recognized the existence of difficult trade-offs.

I believe success will depend on multidisciplinary programmes embracing an even wider span of disciplines across the natural and social sciences than characterized the Green Revolution. The teams will need to be equipped not only with the latest technical tools, but also with powerful methods of interdisciplinary analysis, such as Agroecosystem Analysis, which I described in Chapter 10. One of the most difficult challenges will be achieving parity of esteem among research and development scientists.

Agricultural science has been long dominated by the biological scientists, who have set priorities and often viewed their social-science colleagues – economists, sociologists, anthropologists – as there to provide useful advice on how to introduce new varieties or technologies, and to identify constraints where implementation fails. There is an almost universal tendency to create hierarchies which elevate the 'hard' sciences above the 'soft', and within the 'hard' give pre-eminence to those working at the cellular and molecular level. It is a tendency that is likely to be reinforced with the growing importance of the science and technology underlying genetic engineering. Yet if the objective of the Doubly Green Revolution is to ensure that the poor have access to the food they require, the social scientists are going to have an equally important role in analysing the needs and priorities of the poor and in helping to translate them into research agendas.

In writing this I do not intend to understate the great potential of the recent revolutions in molecular biology and ecology. But, I maintain, we face an equally important challenge in providing the conditions under which this potential can be realized. Because of the complexity of the science and the range of skills and experience required, few developing countries can claim self-sufficiency in science and technology. Even many of the most developed nations are relying on international collaboration. At the other end of the spectrum, countries worst affected by hunger and poverty frequently lack sufficient resources to deal with relatively straightforward problems. For some time to come, international research is going to be crucial – providing the developing countries with access to new knowledge and technology and helping to strengthen their research capacities. And wherever research is likely to benefit from significant economics of scale or scope, collaboration will be needed, particularly at a regional level.

The national agricultural research systems (NARS) of the developing countries include some institutions of international stature, but many are weak and under-resourced. Although new public agricultural research institutes and universities were created in the 1970s and early 80s, by the end of the decade public deficits were producing cutbacks. Some NARS have suffered major crises, the most severe in Sub-Saharan Africa, where investment in agricultural research has hardly grown. The notable exception has been China, where investment exceeds that of the whole of the rest of Asia (Figure 16.1). One solution to the crisis has been privatization but,

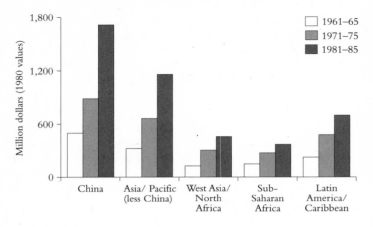

Figure 16.1 Annual agricultural research expenditures [8]

in practice, this has proved difficult to achieve. There has also been slow progress in setting up farmer-supported research organizations. Producer cooperatives, in various guises, have proved successful at organization of inputs and marketing, but few have extended their remit to research. Success has been more apparent in the research programmes funded and supported by national and international non-governmental organizations (NGOs) [9].

Despite the cutbacks in aid, the developed countries have continued to provide support to NARS through collaborative links, involving universities and research centres. Several European countries have maintained public agencies for international agricultural research and have recently set up the European Consortium for Agricultural Research in the Tropics (ECART). The United States has comparable programmes, often involving the Land Grant universities, and funded by the US Agency for International Development (USAID). Ford, Rockefeller and other private foundations continue to be involved in the funding of the International Agricultural Research Centres. Japan, Canada and Australia have set up institutions specializing in scientific cooperation. The Food and Agriculture Organization has recently established a programme in sustainable agriculture. But these efforts are not well coordinated. Information circulates poorly and there is insufficient interaction between the different agencies.

These inter-agency deficiencies must be remedied, not least because many of the issues and problems we face involve important international externalities [10]. In an increasingly close-knit world the actions of one country frequently affect the well-being of another. The effects, of course, may be negative or positive. Major irrigation schemes can have deleterious consequences on downstream countries, upland reforestation can have beneficial effects. But because the costs and benefits of such externalities are not fully borne by the country responsible, it is likely to under- or over-invest in the appropriate research and development. Because they will not capture all the benefits from the research, countries are unlikely to devote enough research to:

- Genetically engineered crops that can be used in other countries;
- Approaches to reducing carbon emissions (e.g. sustainable forestry);
- The conservation of biodiversity; or
- Technologies to prevent soil erosion affecting international waterways –

Similarly, countries are likely to over-invest in research activities that indirectly promote deforestation, or water pollution, where the environmental costs are borne by other countries. The role of international agricultural and natural-resource research is to help correct for these deficiencies, generating international 'public goods' that provide benefits across national boundaries. Often the problems are common and so are the solutions.

At the core of international research collaboration, in the past and for the foreseeable future, lie the International Agricultural Research Centres (IARCs). When IRRI was founded in 1960 neither Ford nor Rockefeller envisaged setting up similar centres, but IRRI's success encouraged them to repeat the pattern. By 1967 three additional centres had been formally organized, and it was not long before the growing number of centres, despite their independence, needed a more organized system of support that could increase the pool of funding and begin to plan for further expansion. In 1971 the Consultative Group on International Agricultural Research (CGIAR) was created as an informal consortium of the World Bank and major regional banks, various UN agencies and the leading foundations, together with developed- and developing-country representations [11]. By 1996 CGIAR was supporting sixteen centres, employing a total of 1,000 scientific staff, to the tune of $300 million (see Appendix). Compared with the over $9 billion being spent on agriculture research globally, including over $4 billion in the developing countries, this is an exceedingly small sum [12]. But the return has been immense. The CGIAR centres have played a key leadership role, serving as creative conduits between developing-country research systems and advanced institutions in the developed countries. And their political independence, open accountability and emphasis on externally monitored excellence in research have contributed, as I described in Chapter 4, to a remarkable record of research achievements over the past thirty years – in germ-plasm characterization, plant-breeding, pathology, pest control, crop–livestock and agroforestry systems and the field application of new technologies of tillage and soil conservation.

For the future we shall need many more such achievements. But, as I have argued, there will need to be a significant shift in emphasis that recognizes:

– The goal of realizing high, sustainable productivity in both high- and lower-potential areas, along with greater employment and incomes so that the poor have access to the food they require;
– A need to foster research partnerships that go beyond a simple transfer of technology, extending from socio-economic need to technology and back again;
– The opportunity to exploit, in an interdisciplinary and international fashion, the new paradigms of molecular biology and ecology;
– The research requirements of a wide range of agroecosystems involving greater in-country expertise and farmer participation in research; and
– The need to work to agendas of research outputs that cross country boundaries.

There is a continuing role, here, for the CGIAR and its centres in helping to shape the programme and to create the necessary partnerships. The centres have already made considerable progress in developing the new agenda, partly through a reorientation of mission in the older centres. Far more emphasis is now being placed on breeding varieties for the rain-fed and semi-arid lands, on a wider range of crops and on integrated approaches to pest and disease control and nutrient management. The research priorities are also being strongly influenced by the newer centres admitted to the CGIAR in 1989, some, such as the International Centre for Research in Agroforestry (ICRAF), with a natural–resource orienta-

---

Box 16.2 **System–wide initiatives being developed under the CGIAR** [13]

- System–wide genetic resources
- Indo–Gangetic plains rice–wheat
- Latin American hillsides programme
- Alternatives to slash-and-burn
- Sustainable mountain agricultural development
- On-farm water-husbandry in West Africa/North Africa
- Global livestock programme
- Water management
- Agricultural research indicators
- Property rights and collective action
- Soil, water and nutrient management
- Integrated pest management

---

tion, and others, such as the International Irrigation Management Institute (IIMI), with social-science strengths.

In 1995 the Vision Panel of the CGIAR [14] set out a plan for the future based on two types of programmes:

– Global programmes, determined internationally and geared toward strategic research problems of cross-regional significance. These would focus on the development of genetic materials for selected crop, livestock, forestry, and fish species that are recognized as keys to sustainable food security and on such problems as agriculture's contributions to global pollution. They would also include a number of system–wide programmes of research and development (Box 16.2).
– Regional action programmes, determined by groups of countries, which address specific sustainable-production problems faced in geographic regions where increased production is needed most urgently, particularly in Sub-Saharan Africa and South Asia, and especially those areas of lower agricultural potential where the poorest live [15].

Both kinds of programme will involve partnerships, along the lines of those I described in Chapter 8, where private industry, the IARCs and developing-country research institutes have come together to exploit their comparative advantages in genetic engineering (Box 16.3).

Box 16.3 **The participants in potential partnerships in international agricultural research**

Industrialized countries
Research institutes
Universities
Private companies
Consortia

Consultative Group on International Agricultural Research (CGIAR)
The International Agricultural Research Centres (IARCs)

Developing countries
(The National Agricultural Research Systems, NARS)
Regional research institutes
National research institutes
Universities
Private companies
Non-government organizations
Farmers

This new approach will involve a significant change in the manner of funding and the relations between the centres, their donors and the NARS of the developing countries. It will also require a shift in the style of research, particularly for the regional action programmes. No longer should research begin or end on research stations. Moreover, it will not be enough for the research teams to be confined to scientists and extension workers alone. As I argued in Chapter 10, the new goals require that farmers and rural communities are intimately involved in the processes of analysis, design and experimentation, and in the eventual application of technology. The new research priorities must be rooted in an intimate understanding of the needs of poor farm households.

In recent years there has been a growing involvement of NGOs in agricultural research and development. A comprehensive review by John Farrington and his colleagues of the Overseas Development Institute, based on seventy case studies, has described activities which range from the promoting of farmer organizations, training activities, the elaboration of methods of diagnosis and technology development to innovation and its dissemination [16]. NGOs have considerable comparative advantage in fostering participatory research and development. They have close knowledge of the reality of poor people's livelihoods, especially where the NGO has strong local roots, established over several years. In particular, NGOs can place agricultural innovation within the broader context of livelihoods, and are sensitive to the dynamics of households and the wider community. As the survey has shown they are playing a major role in introducing government agencies to participatory methods and insights.

However, NGOs often lack technical knowledge and have a limited capacity to draw on some of the latest scientific developments. It is in this respect that the International Agricultural Research Centres have a comparative advantage. I have already described, in Chapter 10, some of the scientist—farmer partnerships created by CIAT, and most of the other centres have been similarly active. However, many of those individuals in the IARCs who have pioneered these approaches have tended to be isolated and marginalized in their institutions. The hierarchy I mentioned earlier has worked against them. There is also a concern, particularly at institutes such as IRRI and CIMMYT, that some of the pressure to move more 'upstream' and concentrate on strategic research, leaving applied research to the National Agricultural Research Systems, will mean the end of scientist—farmer partnerships, just as they are beginning to pay off, not only in terms of benefits to farmers but in creating a greater understanding among scientists of the realities of poor farm livelihoods.

At the heart of the Doubly Green Revolution is not simply a new set of scientific knowledge and techniques, nor even new approaches to participation, but a change in attitude and perception [17]. In many respects it goes against long-held beliefs in research and extension institutions. Individuals who have received many years of training and value their expert knowledge have to come to see a role for farmers' own knowledge and perceptions. This is not easy. Although those who participate, on both sides, soon appreciate the benefits, the initial steps are large and require support and encouragement, with the active help of those who have experience in the process. For the future, this is going to need continued activity on the part of all agricultural development agencies – IARCs, NARS, NGOs and government agencies – and financial support from both developing-country governments and the international donor community.

315

In this book I have tried to describe the scale of the problems we face. There are currently some three-quarters of a billion people chronically undernourished and the best predictions suggest this number will change little over the next twenty-five years. Declines in yield growth, threats to sustainability, the lack of appropriate research capability and insufficient attention to the need to create rural employment and incomes cast doubts on whether the goal of the World Food Summit is attainable. The private sector has an important role to play, freer trade will eventually provide incentives and encourage efficiency but market forces, for the reasons I have articulated, are not going to be enough. If we are to eradicate hunger early in the next century it will require a concerted effort of publicly funded research and development.

In the latter part of the book I have been deliberately somewhat upbeat, partly because my experience leads me to believe the challenge can be met successfully. I have described technologies that are revolutionary, approaches which promise new ways of tackling problems and a great array of methods and techniques developed by scientists and farmers alike that hold promise for solving particular problems. But, in concluding, I need to stress how little we still know and the downsides of what I have been proposing.

We still have severe gaps in our knowledge. We need better statistical data on the incidence of and the changes in poverty and hunger. Where is the situation improving and why? We need more insights into the nature and dynamics of poor households and their livelihoods. In specific situations, how can we use agricultural and natural-resource development to increase employment and incomes? Scientists need to provide a better understanding of the genetic basis of yield, of such key processes as photosynthesis and nitrogen fixation, and of the responses of plants to stress, and to show how this knowledge can be exploited in conventional plant-breeding and genetic engineering. We need greater understanding not only of the ecological underpinnings of integrated pest, disease and nutrient management, but also of the economic and institutional require-ments for success. And this is equally true of how we manage our natural resources: rangelands, forests and fisheries.

I have repeatedly stressed the importance of a participatory approach and described the many successful examples of programmes in which farmers have played a key role in the research and development programme. I have highlighted the revolutionary nature of the new participatory techniques, which can give the poor a voice and power, and change the attitudes and agendas of research and development experts. But, as a reading of the literature reveals, the successes are not easily achieved, failures occur and, as programmes develop, limitations of knowledge, or of technologies or institutions, occur. Most important of all: however good the technologies, however well thought out the development programme, however committed and well trained the programme staff, the larger environment and particularly the prevalent social and economic policies can be the overriding determinants of success and failure.

These questions are raised and discussed more fully in the literature I have cited in the previous chapters. How the knowledge can be acquired and the problems of the new approaches overcome is described elsewhere, and in more detail. This book is partly an argument and partly an introduction. More important, it is a call for action – to donors, to policy-makers, to scientists, to development workers in research institutes and NGOs, and to people anywhere without special expertise but with a concern for the future of the world. Now is not the time to sit back and congratulate ourselves on what has been achieved over the past thirty years. It is the next thirty years that will be the true test of whether we can harness the power of science and technology not just for the better-off, or even the majority, but for those millions of poor and hungry who deserve and have a right to enough to eat.

Notes

[1] FAO, 1996, *Rome Declaration on World Food Security and World Food Summit Plan of Action*, World Food Summit, 13–17 November 1996, Rome (Italy), Food and Agriculture Organization

[2] G. Gordillo de Anda and R. Trenchard (in press), 'World Food Summit: forging a new covenant for the new millennium', in J. C. Waterlow, D. G. Armstrong, L. Fowden and R. Riley (eds.), *Feeding a World Population of More than Eight Billion People: A Challenge to Science*, Oxford, Oxford University Press

[3] Gordillo de Anda and Trenchard, op. cit.

[4] FAO, 1996, op. cit.

[5] FAO, 1993, *Food Aid in Figures*, Rome, Food and Agriculture Organization; FAOSTAT TS: SOFA '95, Rome, Food and Agriculture Organization (diskette)

[6] G. R. Conway, U. Lele, J. Peacock and M. Piñeiro, 1994, *Sustainable Agriculture for a Food Secure World*, Washington DC, Consultative Group on International

Agricultural Research/Stockholm (Sweden), Swedish Agency for Research Cooperation with Developing Countries

[7] FAO, 1996, op. cit.

[8] J. R. Anderson, P. G. Pardey and J. Roseboom, 1994, 'Sustaining growth in agriculture. A quantitative review of agricultural research investments', *Agricultural Economics*, 10, 107–23

[9] J. Farrington and A. Bebbington, 1994, *From Research to Innovation: Getting the Most from Interaction with NGOs in Farming Systems Research and Extension*, London, International Institute for Environment and Development (Gatekeeper Series No. 43); J. Farrington and A. J. Bebbington, 1993, *Reluctant Partners? Non-Governmental Organisations, the State and Sustainable Agricultural Development*, London, Routledge

[10] Conway *et al.*, op. cit.

[11] W. C. Baum, 1986, *Partners against Hunger: The Consultative Group on International Agricultural Research*, Washington DC, World Bank

[12] Annual expenditures 1981–5 in Anderson *et al., op. cit.*

[13] D. J. Greenland, 1996, 'International agricultural research and the CGIAR System – past, present and future', Reading (UK), University of Reading (mimeo.)

[14] Conway *et al.*, op. cit.

[15] IITA, 1992, *Sustainable Food Production in Sub-Saharan Africa. 1. IITA's Contribution*, Ibadan (Nigeria), International Institute of Tropical Agriculture

[16] J. Bebbington, M. Prager, H. Riveros and G. Thiele, 1993, *NGOs and the State in Latin America: Rethinking Roles in Sustainable Agricultural Development*, London, Routledge; Farrington and Bebbington, 1994, op. cit.; J. Farrington and D. Lewis (eds.), 1993, *NGOs and the State in Asia: Rethinking Roles in Sustainable Agricultural Development*, London, Routledge; K. Wellard and J. G. Copestake (eds.), 1993, *Non-Governmental Organisations and the State in Africa: Rethinking Roles in Sustainable Agricultural Development*, London, Routledge

[17] R. Chambers, 1997, *Whose Reality Counts? Putting the Last First*, London, Intermediate Technology

Appendix: International Agricultural Research Centres*

Centro Internacional de Agricultura Tropical (CIAT)

(International Centre for Tropical Agriculture)

Headquarters: Cali (Colombia). Founded 1967

Focus: To contribute to the alleviation of hunger and poverty in tropical countries by applying science to the generation of technology that will lead to lasting increases in agricultural output while preserving the natural resource base. Research is conducted on germ-plasm development of beans, cassava, tropical forages, and rice for Latin America and on resource management in humid agroecosystems in tropical America, including hillsides, forest margins and savannas.

Center for International Forestry Research (CIFOR)

Headquarters: Bogor (Indonesia). Founded 1992

Focus: To contribute to the sustained well-being of people in developing countries, particularly in the tropics, through collaborative strategic and applied research in forest systems and forestry, and by promoting the transfer of appropriate new technologies and the adoption of new methods of social organization for national development.

Centro Internacional de Mejoramiento de Maíz Trigo (CIMMYT)

(International Centre for the Improvement of Maize and Wheat)

Headquarters: Mexico City (Mexico). Founded 1966

Focus: To help the poor by increasing the productivity of resources committed to maize and wheat in developing countries, while protecting the environment, through agricultural research and in concert with national research systems.

* Consultative Group on International Agricultural Research (CGIAR) centres.

Centro Internacional de la Papa (CIP)

(International Potato Center)

Headquarters: Lima (Peru). Founded 1971

Focus: To contribute to increased food production, the generation of sustainable and environmentally sensitive agricultural systems, and improved human welfare by conducting coordinated, multidisciplinary research programmes on potato and sweet potato, by carrying out worldwide collaborative research and training, by catalysing collaboration among countries in solving common problems, and by helping scientists worldwide to respond flexibly and successfully to changing demands in agriculture.

International Center for Agricultural Research in the Dry Areas (ICARDA)

Headquarters: Aleppo (Syria). Founded 1977

Focus: To meet the challenge posed by a harsh, stressful and variable environment in which the productivity of winter rain-fed agricultural systems must be increased to higher sustainable levels, in which soil degradation must be arrested and possibly reversed, and in which water-use efficiency and the quality of the fragile environment need to be ensured. ICARDA has a world responsibility for the improvement of barley, lentils and faba bean, and a regional responsibility in West Asia and North Africa for the improvement of wheat, chickpea, forages and pasture. ICARDA emphasizes rangeland improvement, small ruminant management and nutrition, and rain-fed farming systems associated with these crops.

International Center for Living Aquatic Resources Management (ICLARM)

Headquarters: Metro Manila (Philippines). Founded 1977

Focus: To improve the production and management of aquatic resources, for sustainable benefits to present and future generations of low-income producers and consumers in developing countries, through international multidisciplinary research in partnership with national agricultural research systems. The declining state and threatened sustainability of fisheries caused by overfishing exacerbated with poverty and pollution, and the potential for increases in aquaculture production, call for research which includes understanding of the dynamics of coastal and coral-reef resource systems and of integrated agriculture–aquaculture systems, investigating alternative management schemes in these systems, and improving the productivity of key species.

International Centre for Research in Agroforestry (ICRAF)

Headquarters: Nairobi (Kenya). Founded 1977

Focus: To mitigate tropical deforestation, land depletion and rural poverty through improved agroforestry systems. Trees in farming systems can increase and diversify farmer income, make farming systems more robust, reverse land degradation, and reduce the pressure on natural forests. ICRAF carries out research with national agricultural and forestry research systems, non-governmental organizations, and other research partners, and is focused on two major thrusts: finding alternatives to slash-and-burn agriculture in the humid tropics; and overcoming land depletion in subhumid and semi-arid Africa.

International Crops Research Institute for the Semi-Arid Tropics (ICRISAT)

Headquarters: Patancheru, Andhra Pradesh (India). Founded 1972

Focus: To conduct research leading to enhanced sustainable food production in the harsh conditions of the semi-arid tropics. ICRISAT's main crops – sorghum, finger millet, pearl millet, chickpea, pigeonpea, and groundnut – are not generally known in the world's more favourable agricultural regions, but they are vital to life for the one-sixth of the world's population that lives in the semi-arid tropics. ICRISAT conducts research in partnership with the national agricultural systems that encompasses the management of the region's limited natural resources to increase the productivity, stability and sustainability of these and other crops.

International Food Policy Research Institute (IFPRI)

Headquarters: Washington DC (USA). Founded 1975

Focus: IFPRI was established to identify and analyse alternative national and international strategies and policies for meeting the food needs of the developing world on a sustainable basis, with particular emphasis on low-income countries and on the poorer groups in those countries. While IFPRI's research is specifically geared to contributing to the reduction of hunger and malnutrition, the factors involved are many and wide-ranging, requiring analysis of underlying processes and extending beyond a narrowly defined food sector. IFPRI collaborates with governments and private and public institutions worldwide interested in increasing food production and improving the equity of its distribution. Research results are disseminated to policy-makers, administrators, policy analysts, researchers and others concerned with national and international food and agricultural policy.

International Irrigation Management Institute (IIMI)

Headquarters: Colombo (Sri Lanka). Founded 1984

Focus: IIMI's mission is to foster improvement in the management of water-resource systems and irrigated agriculture. IIMI conducts a worldwide programme to generate knowledge to improve water-resource capacity, and to support the introduction of improved technology, policies and management approaches.

International Institute of Tropical Agriculture (IITA)

Headquarters: Ibadan (Nigeria). Founded 1967

Focus: IITA conducts research and outreach activities, with partner programmes in countries of Sub-Saharan Africa, to help those countries increase food production on an ecologically sustainable basis. IITA seeks to improve the food quality, plant health and post-harvest processing of its mandated crops – cassava, maize, cowpea, soybean, yam, and banana and plantain – while strengthening national research capabilities.

International Livestock Research Institute (ILRI)

Headquarters: Nairobi (Kenya). Founded 1995

Focus: To increase animal health, nutrition, and productivity (i.e. milk, meat, traction) by removing constraints on tropical livestock production, particularly among small-scale farmers; to protect environments supporting animal production against degradation by tailoring production systems and developing technologies that are sustainable over the long term; to characterize and conserve the genetic diversity of indigenous tropical forage species and livestock breeds; and to promote equitable and sustainable national policies for the development of animal agriculture and the management of natural resources affected by animal production, encouraging, in particular, those policies that support strategies for reducing hunger and poverty, for improving food security, and for protecting the environment.

International Plant Genetic Resources Institute (IPGRI)

Headquarters: Rome (Italy). Founded 1974

Focus: To encourage, support and engage in activities to strengthen the conservation and use of plant genetic resources worldwide, with special emphasis on developing countries, by undertaking research and training and by providing scientific and technical information.

International Rice Research Institute (IRRI)

Headquarters: Manila (Philippines). Founded 1960

Focus: To improve the well-being of present and future generations of rice farmers and consumers, particularly those with low incomes, by generating and disseminating rice-related knowledge and technology of short- and long-term environmental, social and economic benefits and by helping to enhance national rice research.

International Service for National Agricultural Research (ISNAR)

Headquarters: The Hague (Netherlands). Founded 1979

Focus: To help developing countries bring about sustained improvements in the performance of their national agricultural research systems and organizations. ISNAR does this by supporting their efforts in institutional development, promoting appropriate policies and funding for agricultural research, developing or adapting improved research management and techniques, and generating and disseminating relevant knowledge and information.

West Africa Rice Development Association (WARDA)

Headquarters: Bouaké (Côte d'Ivoire). Founded 1970

Focus: WARDA's work is aimed at strengthening the capability of agricultural scientists in West Africa for technology generation to increase the sustainable productivity of intensified rice-based cropping systems in a manner that improves the well-being of resource-poor farm families and that conserves and enhances the natural resource base. Research covers rice grown in mangrove swamps, inland valleys, upland conditions and irrigated conditions.

Index